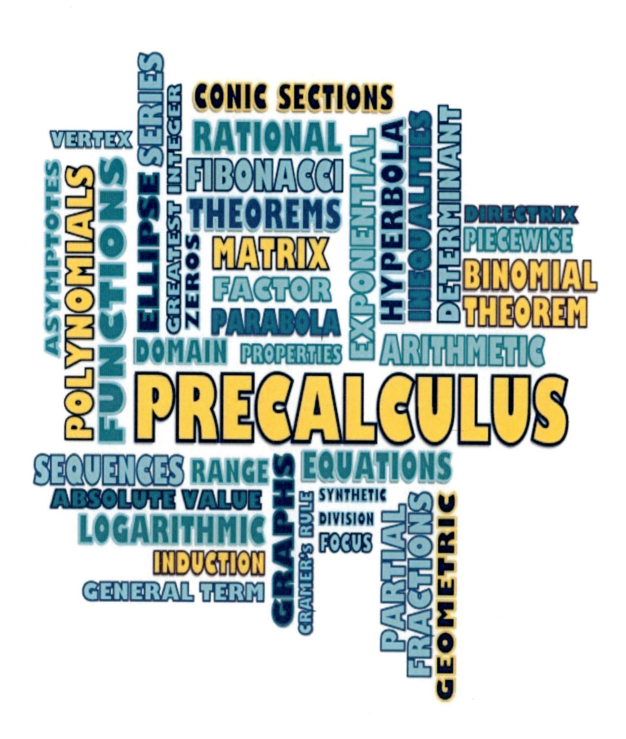

PRECALCULUS, Second Edition
Copyright 2017

Terri Bedford, Author
Associate Professor of Mathematics
Gulf Coast State College

Michael Brinegar, Editor
Associate Professor of Mathematics
Gulf Coast State College

Special thank you to:
Gulf Coast State College Board of Trustees
Dr. John Holdnak, President
Dr. Holly Kuehner, Vice-President of Academic Affairs
Angelia Reynolds, Chair of Mathematics
Scott Spencer, Senior Administrative Assistant, Mathematics
Leslie Hapner, Dean of Business Affairs

Table of Contents

Lesson 1	Piecewise Functions	Page 5
Lesson 2	Synthetic Division	Page 20
Lesson 3	Zeros of a Polynomial	Page 29
Lesson 4	Graphing a Polynomial	Page 39
Cumulative Review 1		Page 51
Lesson 5	Rational Functions	Page 54
Lesson 6	Inequalities	Page 66
Lesson 7	Exponential and Logarithmic Functions	Page 78
Lesson 8	Properties of Logarithms	Page 88
Lesson 9	Exponential and Logarithmic Equations	Page 95
Cumulative Review 2		Page 103
Lesson 10	Introduction to Matrices	Page 106
Lesson 11	Algebra of Matrices	Page 116
Lesson 12	Matrix Inverse	Page 127
Lesson 13	Determinants	Page 136
Lesson 14	Partial Fraction Decomposition	Page 147
Cumulative Review 3		Page 156
Lesson 15	Factoring Completely	Page 158
Lesson 16	Circle and Ellipse	Page 164
Lesson 17	Hyperbola	Page 172
Lesson 18	Parabola	Page 181
Cumulative Review 4		Page 191
Lesson 19	Sequences and Series	Page 193
Lesson 20	Arithmetic	Page 208
Lesson 21	Geometric	Page 216
Lesson 22	Mathematical Induction	Page 229
Lesson 23	Binomial Theorem	Page 243
Cumulative Review 5		Page 249
Algebra Review Solutions		Page 252
Keep It Fresh Solutions		Page 254
Index		Page 258

Algebra Review:

Graphing Basic Functions	Page 5
Factoring Polynomials	Page 13
Polynomial Division	Page 25
Quadratic Equations	Page 25
Complete the Square	Page 45
Distance Formula	Page 45
Midpoint Formula	Page 45
Interval Notation	Page 61
Combine Rational Expressions	Page 73
Composition of Functions	Page 87
Inverse Functions	Page 87
Logistic Growth Model	Page 102
Algebra Review Solutions	Page 252

Lesson 1 Piecewise Functions

I. Review Graphs of Basic Functions

Linear $y = mx + b$

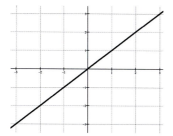

Domain $(-\infty, \infty)$

Range $(-\infty, \infty)$

Absolute Value $y = |x|$

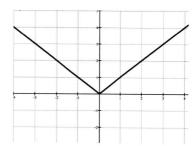

Domain $(-\infty, \infty)$

Range $[0, \infty)$

Quadratic $y = x^2$

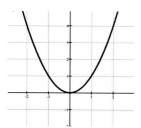

Domain $(-\infty, \infty)$

Range $[0, \infty)$

Square Root $y = \sqrt{x}$

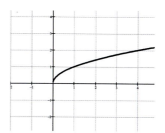

Domain $[0, \infty)$

Range $[0, \infty)$

Cubic $y = x^3$

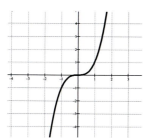

Domain $(-\infty, \infty)$

Range $(-\infty, \infty)$

Rational $y = \dfrac{1}{x}$

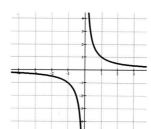

Domain $(-\infty, 0) \cup (0, \infty)$

Range $(-\infty, 0) \cup (0, \infty)$

The following is a list of transformations that were studied in algebra. Knowing the shape of the six basic functions and the transformations will help graph more complicated functions.

Horizontal Shifts	$f(x+c)$	If $c > 0$, shift the graph left c units. If $c < 0$, shift the graph right c units.
Reflections	$f(-x)$ $-f(x)$	Reflect the graph across the y-axis. Reflect the graph across the x-axis.
Vertical Shrink/Stretch	$cf(x)$	If $c > 1$, the graph stretches by a factor of c. If $0 < c < 1$, the graph shrinks by a factor of c.
Horizontal Shrink/Stretch	$f(cx)$	If $c > 1$, the graph shrinks by a factor of c. If $0 < c < 1$, the graph stretches by a factor of c.
Vertical Shifts	$f(x)+c$	If $c > 0$, shift the graph up c units. If $c < 0$, shift the graph down c units.

Example 1: Name the transformations for the function.

$g(x) = -(x+1)^2 + 1$

Solution: The graph of the quadratic function, called a parabola, will have the following transformations.

- Horizontal Shift left 1 unit
- Reflection about the x-axis
- Vertical Shift up 1 unit

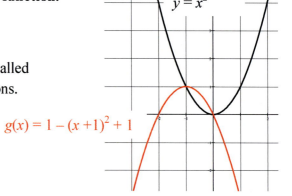

In calculus, continuity will be defined in more detail. For now, we will use continuity to describe a function that can be drawn without picking up your pencil. In other words, a continuous function does not have any gaps or breaks in the graph. Linear, absolute value, quadratic, square root, and cubic are all continuous functions. The rational function is not continuous at $x = 0$.

We use the word smooth to describe a function that does not have any sharp corners or edges. The Absolute Value graph is not smooth, since it has a sharp corner at the vertex point $(0,0)$.

II. Piecewise Functions

A **piecewise function** is a function represented by more than one expression. Each expression has a specified domain that does not overlap with other expressions. These graphs may be various combinations of smooth/not smooth and continuous/discontinuous.

Example 2: Graph the piecewise function.

$$f(x) = \begin{cases} -\dfrac{1}{3}x+4 & \text{for } x \geq 0 \\ -(x+1)^2 +1 & \text{for } x < 0 \end{cases}$$

closed point →

x	f(x)
0	4
3	3
6	2

← open point

x	f(x)
0	0
–1	1
–2	0

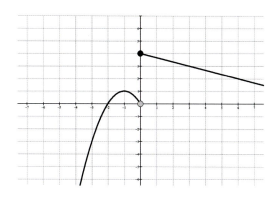

Solution: To graph a piecewise function, start by making a table of values for each piece. Pay special attention to the specified domain. The linear function is graphed for *x* values greater than or equal to 0, and the parabola is graphed for *x* values less than 0. Be sure to include the endpoints, and make note if the point is closed or open. In the first expression, $(0,4)$ will be a closed point on the graph because the domain includes 0. In the second expression, $(0,0)$ will be an open point on the graph because the domain is less than 0. The final graph of the piecewise function passes the vertical line test and is discontinuous at $x = 0$. The domain is $(-\infty, \infty)$, and the range is $(-\infty, 4]$.

Example 3: Graph the piecewise function.

$$g(x) = \begin{cases} 4-\sqrt{x} & \text{for } x > 1 \\ -1 & \text{for } -2 < x \leq 1 \\ 2x+3 & \text{for } x \leq -2 \end{cases}$$

open point →

x	g(x)
9	1
4	2
1	3

closed point ↓

x	g(x)
1	–1
0	–1
–2	–1

closed point ↓

x	g(x)
–2	–1
–3	–3
–4	–5

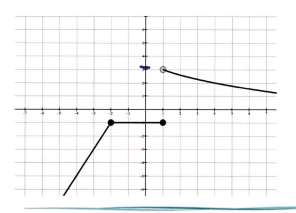

Solution: Make a table of values for each piece. Be sure to include the endpoints and make note if the point is closed or open. For the first expression, $(1,3)$ will be an open point on the graph because the domain is strictly greater than 1. Notice the second expression will have two endpoints, $(-2,-1)$ will be an open point and $(1,-1)$ will be a closed point. For the third expression, $(-2,-1)$ will be a closed point on the graph because the domain includes -2. Did you notice the point $(-2,-1)$ was included in the 2nd and 3rd expressions? The 2nd and 3rd expressions connect at $(-2,-1)$. When this happens, the function is continuous at the point. The closed point is shown on the graph. The piecewise function is continuous at $x = -2$ and discontinuous at $x = 1$. The domain is $(-\infty, \infty)$, and the range is $(-\infty, 3)$.

Example 4 shows another piecewise function that is not continuous. This example has a removable discontinuity, because the function can be made continuous by redefining a single point.

Example 4: Graph the piecewise function.

$$h(x) = \begin{cases} \dfrac{x^2 + 3x - 4}{x + 4} & \text{for } x \neq -4 \\ 0 & \text{for } x = -4 \end{cases}$$

x	h(x)
−6	−7
−5	−6
−4	0
−3	−4
−2	−3
−1	−2

Solution: The domain for each piece is different from the previous two examples. We use the first expression in the function for all x values except for -4. Therefore, our table will look a little different. For $x = -4$, we use the second expression. For the second piece of this function, we will plot only the point $(-4, 0)$.

To make the table for the first expression, notice that we can factor the numerator and reduce the rational function.

$$\frac{x^2 + 3x - 4}{x + 4} = \frac{(x+4)(x-1)}{x+4} = x - 1$$

Using the simplified expression, we see the graph is a line with a hole at $x = -4$. Where is the hole located? We can find the y value using the simplified expression of the fraction. Evaluate $x = -4$ in the expression $x - 1$, and we see the hole is located at the point $(-4, -5)$. The hole is located at the removable discontinuity.

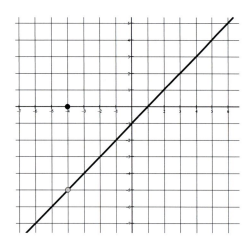

You can find many applications of piecewise functions. For example, rarely can you find just one function to model a situation. You may find the profit of a business grows using a square root function for 10 years, only to find the profit declines using a linear function the next 5 years. Because of this, a piecewise function is needed to represent the data over the 15 year span.

In some cases, decimal values are not used in computations. For example, if you make a phone call that lasts for 5.3 minutes, you are charged for 6 minutes. The same would be true for an 8.1 minute phone call; you are charged for 9 minutes.

Take a look at bank amounts. If you earn interest on an account, you are not given decimal cents. For example, if you earn $56.876, your bank gives you $56.87. If your account earned $4.483 cents, you are given $4.48. The partial, fraction part of the cents is not accumulated. This is an example of the **Greatest Integer Function**. The Greatest Integer Function is a piecewise function that gives the largest integer less than or equal to x. There are several different notations, and you can also find several other names used to describe the function.

$[\![x]\!]$ or $\lfloor x \rfloor$ or $\text{int}(x)$ "floor function" or "step function"

Example 5: Graph $f(x) = [\![x]\!]$.

Solution: First, we need to evaluate the function for a variety of x values to determine a pattern. Let's take a look at values between 2 and 3.

$f(2.105) = [\![2.105]\!] = 2$
$f(2.45) = [\![2.45]\!] = 2$
$f(2.668) = [\![2.668]\!] = 2$
$f(2.993) = [\![2.993]\!] = 2$

For any x value between 2 and 3, the function results in the answer 2. For $x = 2$ and $x = 3$, the function gives $f(2) = [\![2]\!] = 2$ and $f(3) = [\![3]\!] = 3$.

In other words, $f(x) = 2$ for $2 \le x < 3$ and $f(x) = 3$ for $3 \le x < 4$.

Take a look at some negative values as well.
Can you see a pattern for values between -3 and -2?
$f(-2.37) = [\![-2.37]\!] = -3$
$f(-2.412) = [\![-2.412]\!] = -3$
$f(-2.5) = [\![-2.5]\!] = -3$
$f(-2.90) = [\![-2.90]\!] = -3$

Continue with this pattern for all values, and plot the points. The graph of the Greatest Integer Function resembles steps. Note the closed and open points at the integers.

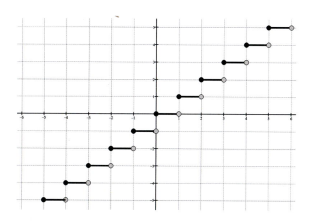

Be careful not to use rounding rules with the Greatest Integer Function. Notice, $[\![2.993]\!] \ne 3$. The Greatest Integer Function is a piecewise function that gives the largest integer less than or equal to x.

9

The same transformations can be used to graph the Greatest Integer Function.

Vertical Shifts	**If $c > 0$, shift the graph up c units.**
$g(x) = [\![x]\!] + c$	**If $c < 0$, shift the graph down c units.**
Horizontal Shifts	**If $c > 0$, shift the graph left c units.**
$g(x) = [\![x + c]\!]$	**If $c < 0$, shift the graph right c units.**
Reflections	
$g(x) = [\![-x]\!]$	**Reflect the graph across the y-axis.**
$g(x) = -[\![x]\!]$	**Reflect the graph across the x-axis.**
Vertical Shrink/Stretch	**If $c > 1$, the graph stretches by a factor of c.**
$g(x) = c[\![x]\!]$	**If $0 < c < 1$, the graph shrinks by a factor of c.**
Horizontal Shrink/Stretch	**If $c > 1$, the graph shrinks by a factor of c.**
$g(x) = [\![cx]\!]$	**If $0 < c < 1$, the graph stretches by a factor of c.**

Example 6: Evaluate the function for the following values. Then use transformations to graph the function. $g(x) = [\![x+3]\!] - 2$

a. $g(-3.91)$ b. $g(-2.54)$ c. $g(4.01)$

Solutions:

a. $g(-3.91) = [\![-3.91+3]\!] - 2$
$= [\![-0.91+3]\!] - 2$
$= -1 - 2$
$= -3$

b. $g(-2.54) = [\![-2.54+3]\!] - 2$
$= [\![0.54]\!] - 2$
$= 0 - 2$
$= -2$

c. $g(4.01) = [\![4.01+3]\!] - 2$
$= [\![7.01]\!] - 2$
$= 7 - 2$
$= 5$

The Greatest Integer Function will shift left 3 units and down 2 units.

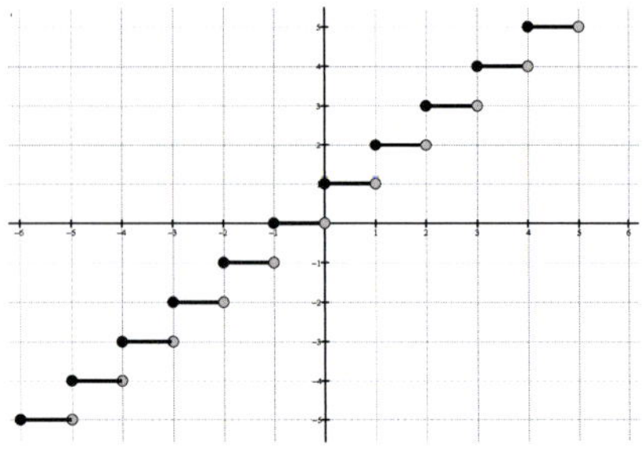

Absolute value is defined as the distance from zero, which can never be negative. The Absolute Value Function $f(x)=|x|$ can be written as a piecewise function. The graph consists of two lines. One line $y=x$ has a positive slope and the other line $y=-x$ has a negative slope.

$$f(x)=|x|=\begin{cases} x & \text{for } x \geq 0 \\ -x & \text{for } x < 0 \end{cases}$$

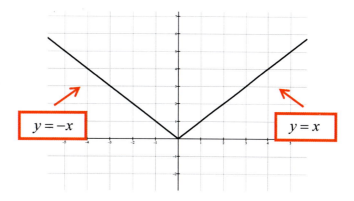

Example 7: Rewrite the absolute value function as a piecewise function. Then graph the function.

a. $g(x)=|2x-1|$ b. $h(x)=|x-1|-|x+5|$

Solution:

a. $g(x)=|2x-1|$

Step 1:
Start with the expression in the absolute value bars.
This is the 1st expression in the piecewise function.

$|2x-1|$
$2x-1$

Step 2:
Find the domain for this expression by solving the inequality.

$2x-1 \geq 0$
$x \geq \dfrac{1}{2}$

Step 3:
Find the opposite of the original expression.
This is the 2nd expression in the piecewise function.

$-(2x-1)=-2x+1$

Step 4:
Find the domain for this expression, by solving the inequality.
Don't forget to switch the inequality sign when dividing by a negative.

$-2x+1 > 0$
$x < \dfrac{1}{2}$

Step 5:
Write the function in piecewise form.

$$g(x)=|2x-1|=\begin{cases} 2x-1 & \text{for } x \geq \dfrac{1}{2} \\ -2x+1 & \text{for } x < \dfrac{1}{2} \end{cases}$$

To graph, use a table of values or use the slope of the line to plot points.

x	$g(x)$
-1	3
0	1
$\frac{1}{2}$	0
1	1
2	3

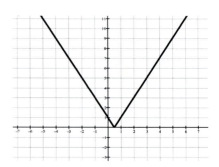

b. $h(x) = |x-1| - |x+5|$

$x-1$

Complete steps 1 – 4 each absolute value expression.

	$|x-1|$	$|x+5|$
Step 1: Start with the expression in the absolute value bars.	$x-1$	$x+5$
Step 2: Find the domain for this expression by solving the inequality.	$x-1 \geq 0$ $x \geq 1$	$x+5 \geq 0$ $x \geq -5$
Step 3: Find the opposite of the original expression.	$-(x-1) = -x+1$	$-(x+5) = -x-5$
Step 4: Find the domain for this expression, by solving the inequality.	$-x+1 > 0$ $x < 1$	$-x-5 > 0$ $x < -5$

To determine the piecewise function, draw a number line to organize the information found in steps 1 – 4. Divide the line into sections using the x values found in steps 2 and 4. Use the inequalities to determine the expression to use in each interval.

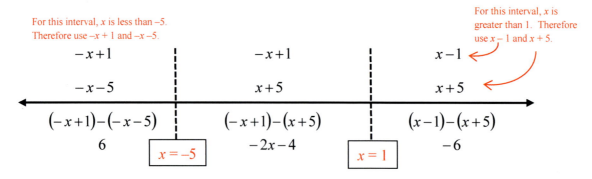

For this interval, x is less than -5. Therefore use $-x+1$ and $-x-5$.

For this interval, x is greater than 1. Therefore use $x-1$ and $x+5$.

Combine using subtraction, and the result is three expressions for the piecewise function. From the number line, the domains for each expression can be found. For the first expression, the domain is $x < -5$. The second expression has domain $-5 \leq x < 1$, and the final expression has domain $x \geq 1$.

$$h(x) = |x-1| - |x+5| = \begin{cases} 6 & \text{for } x < -5 \\ -2x - 4 & \text{for } -5 \le x < 1 \\ -6 & \text{for } x \ge 1 \end{cases}$$

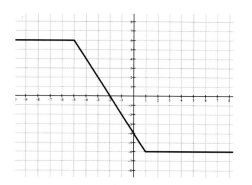

Please note: The functions g(x) and h(x) in Example 7 are continuous. Therefore, the location of the equal to in the domain of the inequality is not important. The important detail to remember is the domains cannot overlap, so you cannot have the equal to in both domains.

OK!

$$g(x) = \begin{cases} 2x - 1 & \text{for } x > \frac{1}{2} \\ -2x + 1 & \text{for } x \le \frac{1}{2} \end{cases}$$

Not OK!

$$g(x) = \begin{cases} 2x - 1 & \text{for } x \ge \frac{1}{2} \\ -2x + 1 & \text{for } x \le \frac{1}{2} \end{cases}$$

$$h(x) = \begin{cases} 6 & \text{for } x < -5 \\ -2x - 4 & \text{for } -5 \le x \le 1 \\ -6 & \text{for } x > 1 \end{cases}$$

$$h(x) = \begin{cases} 6 & \text{for } x < -5 \\ -2x - 4 & \text{for } -5 \le x \le 1 \\ -6 & \text{for } x \ge 1 \end{cases}$$

ALGEBRA REVIEW

Factor the following expressions.

1. $12x^2 + 34x + 24$
2. $6x^3 - 6x^2 - 120x$
3. $25x^2 - 70xy + 49y^2$
4. $x^4 - 13x^2 + 36$
5. $3x^4 - 192$
6. $3x^3 - x^2 + 12x - 4$
7. $54x^3 - 128y^3$
8. $8x^3 + 125y^3$
9. $x^{7/3} - x^{1/3}$
10. $15x^{-1/2} + 5x^{1/2}$

Lesson 1   Practice Exercises

In Exercises 1 – 16, evaluate the function for the specified values. Graph the function and determine the domain and range. Find the values for which the function is discontinuous.

1. $f(x) = \begin{cases} \frac{1}{3}x + 5 & \text{for } x < 0 \\ -4x + 1 & \text{for } x \geq 0 \end{cases}$

   $f(0) = \underline{1}$ and $f(-2) = \underline{4.3}$

2. $g(t) = \begin{cases} t + 4 & \text{for } t \leq 1 \\ 8 - t & \text{for } t > 1 \end{cases}$

   $g(1) = \underline{5}$ and $g(-1) = \underline{3}$

3. $h(x) = \begin{cases} \sqrt{x-2} & \text{for } 2 < x < 6 \\ -3 & \text{for } x \leq 2 \end{cases}$

   $h(-6) = \underline{-3}$ and $h(6) = \underline{und.}$

4. $f(x) = \begin{cases} 4 & \text{for } 0 < x \leq 2 \\ \frac{2}{3}x + 4 & \text{for } x \leq 0 \end{cases}$

   $f(3) = \underline{und.}$ and $f(-1) = \underline{3\frac{1}{3}}$

5. $s(t) = \begin{cases} t^2 - 1 & \text{for } -2 < t < 1 \\ 2 & \text{for } t \geq 1 \end{cases}$

   $s(1) = \underline{2}$ and $s(4) = \underline{2}$

6. $g(x) = \begin{cases} 3x + 2 & \text{for } x > 2 \\ 9 - x^2 & \text{for } -4 < x \leq 2 \end{cases}$

   $g(3) = \underline{11}$ and $g(-6) = \underline{und.}$

7. $h(x) = \begin{cases} 5 - \frac{1}{2}x & \text{for } x \leq 0 \\ \sqrt{x} & \text{for } 0 < x < 4 \\ -1 & \text{for } x \geq 4 \end{cases}$

   $h(-1) = \underline{4\frac{1}{2}\ 5.5}$ and $h(4) = \underline{-1}$

8. $f(x) = \begin{cases} (x+1)^2 - 1 & \text{for } x \leq -1 \\ x & \text{for } -1 < x \leq 1 \\ 1 + \sqrt{x} & \text{for } x > 1 \end{cases}$

   $f(3) = \underline{1 + \sqrt{3}}$ and $f(-1) = \underline{-1}$

9. $f(x) = \begin{cases} x + 3 & \text{for } x \leq -2 \\ \sqrt{4 - x^2} & \text{for } -2 < x < 2 \\ -x + 3 & \text{for } x \geq 2 \end{cases}$

   $f(-2) = \underline{1}$ and $f(0) = \underline{2}$

10. $h(x) = \begin{cases} -x - 3 & \text{for } x < -3 \\ \sqrt{9 - x^2} & \text{for } -3 \leq x < 3 \\ 2x - 2 & \text{for } x \geq 3 \end{cases}$

    $h(-5) = \underline{2}$ and $h(3) = \underline{4}$

11. $h(x) = \begin{cases} \dfrac{x^2-16}{x+4} & \text{for } x \neq -4 \\ 6 & \text{for } x = -4 \end{cases}$

$\dfrac{-15}{5}$

$h(-4) = \underline{6}$ and $h(1) = \underline{-3}$

12. $g(x) = \begin{cases} \dfrac{x^2-25}{x-5} & \text{for } x \neq 5 \\ 10 & \text{for } x = 5 \end{cases}$

$\dfrac{-21}{-7}$

$g(-2) = \underline{3}$ and $g(5) = \underline{10}$

13. $f(x) = \begin{cases} \dfrac{x^3+27}{x+3} & \text{for } x \neq -3 \\ 7 & \text{for } x = -3 \end{cases}$

$f(-3) = \underline{7}$ and $f(0) = \underline{9}$

14. $g(x) = \begin{cases} \dfrac{x^3-8}{x-2} & \text{for } x \neq 2 \\ -5 & \text{for } x = 2 \end{cases}$

$\dfrac{-224}{-8}$

$g(-6) = \underline{28}$ and $g(2) = \underline{-5}$

15. $f(x) = \begin{cases} \dfrac{x^2+3x+2}{x+1} & \text{for } x \neq -1 \\ 1 & \text{for } x = -1 \end{cases}$

$\dfrac{30}{5}$

$f(-1) = \underline{1}$ and $f(4) = \underline{6}$

16. $h(x) = \begin{cases} \dfrac{x^3-1}{x-1} & \text{for } x \neq 1 \\ 3 & \text{for } x = 1 \end{cases}$

$h(0) = \underline{1}$ and $h(1) = \underline{3}$

In Exercises 17 – 22, begin by graphing the Greatest Integer Function, $f(x) = [\![x]\!]$. Then use transformations to graph the given function. Evaluate each function for $x = -4.2$ and $x = 7.6$.

17. $g(x) = [\![x]\!] + 2$ 18. $g(x) = [\![x]\!] - 3$ 19. $g(x) = [\![x+1]\!]$

20. $g(x) = [\![x-4]\!]$ 21. $g(x) = 2[\![x]\!] - 1$ 22. $g(x) = \dfrac{1}{2}[\![x+3]\!]$

In Exercises 23 – 28, write the absolute value function as a piecewise function. Then graph the function.

23. $f(x) = |x-5|$ 24. $g(x) = \left|\dfrac{1}{2}x+1\right|$ 25. $f(x) = |x+2|+|3-x|$

26. $f(x) = |3x-4|+|x+6|$ 27. $f(x) = |5x+1|-|2x-3|$ 28. $f(x) = |x+1|-|x-5|$

15

Solutions to Practice Exercises   **Lesson 1**

1. $f(0) = 1$ and $f(-2) = \dfrac{13}{3}$

   D: $(-\infty, \infty)$ R: $(-\infty, 5)$ Discont at $x = 0$

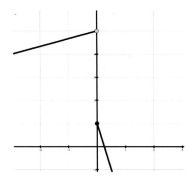

2. $g(1) = 5$ and $g(-1) = 3$

   D: $(-\infty, \infty)$ R: $(-\infty, 7)$ Discont at $t = 1$

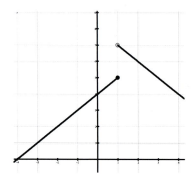

3. $h(-6) = -3$ and $h(6) = $ undefined

   D: $(-\infty, 6)$ R: $\{-3\} \cup (0, 2)$ Discont at $x = 2$

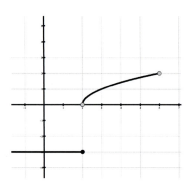

4. $f(3) = $ undefined and $f(-1) = \dfrac{10}{3}$

   D: $(-\infty, 2]$ R: $(-\infty, 4]$ Continuous

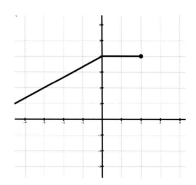

5. $s(1) = 2$ and $s(4) = 2$

   D: $(-2, \infty)$ R: $[-1, 3)$ Discont at $t = 1$

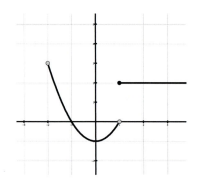

6. $g(3) = 11$ and $g(-6) = $ undefined

   D: $(-4, \infty)$ R: $(-7, \infty)$ Discont at $x = 2$

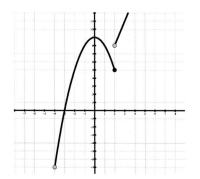

7. $h(-1) = \dfrac{11}{2}$ and $h(4) = -1$

D: $(-\infty, \infty)$ R: $\{-1\} \cup (0,2) \cup [5, \infty)$

Discont at $x = 0$ and 4

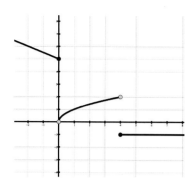

8. $f(3) = 1 + \sqrt{3}$ and $f(-1) = -1$

D: $(-\infty, \infty)$ R: $[-1, \infty)$ Discont at $x = 1$

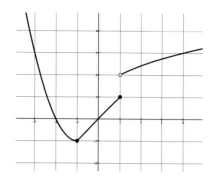

9. $f(-2) = 1$ and $f(0) = 2$

D: $(-\infty, \infty)$ R: $(-\infty, 2]$ Discont at $x = -2$ and 2

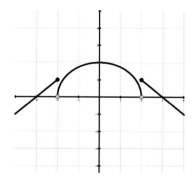

10. $h(-5) = 2$ and $h(3) = 4$

D: $(-\infty, \infty)$ R: $[0, \infty)$ Discont at $x = 3$

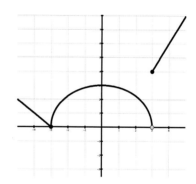

11. $h(-4) = 6$ and $h(1) = -3$

D: $(-\infty, \infty)$ R: $(-\infty, -8) \cup (-8, \infty)$

Discont at $x = -4$

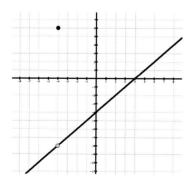

12. $g(-2) = 3$ and $g(5) = 10$

D: $(-\infty, \infty)$ R: $(-\infty, \infty)$ Continuous

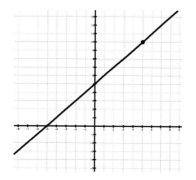

13. $f(-3) = 7$ and $f(0) = 9$

D: $(-\infty, \infty)$ R: $[6.75, \infty)$ Discont at $x = -3$

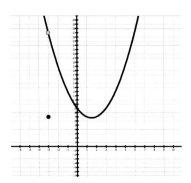

14. $g(-6) = 28$ and $g(2) = -5$

D: $(-\infty, \infty)$ R: $[3, \infty) \cup \{-5\}$ Discont at $x = 2$

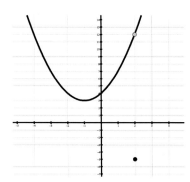

15. $f(-1) = 1$ and $f(4) = 6$

D: $(-\infty, \infty)$ R: $(-\infty, \infty)$ Continuous

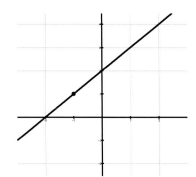

16. $h(0) = 1$ and $h(1) = 3$

D: $(-\infty, \infty)$ R: $\left[\dfrac{3}{4}, \infty\right)$ Continuous

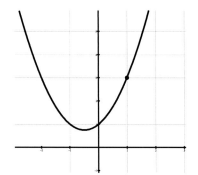

17.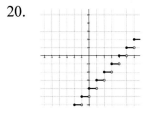

$g(-4.2) = -3; g(7.6) = 9$

18.

$g(-4.2) = -8; g(7.6) = 4$

19.

$g(-4.2) = -4; g(7.6) = 8$

20.

$g(-4.2) = -9; g(7.6) = 3$

21.

$g(-4.2) = -11; g(7.6) = 13$

22.

$g(-4.2) = -1; g(7.6) = 5$

23. $f(x) = \begin{cases} x-5 & \text{for} \quad x \geq 5 \\ -x+5 & \text{for} \quad x < 5 \end{cases}$

24. $g(x) = \begin{cases} \dfrac{1}{2}x+1 & \text{for} \quad x \geq -2 \\ -\dfrac{1}{2}x-1 & \text{for} \quad x < -2 \end{cases}$

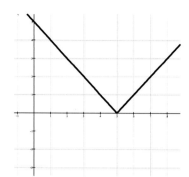
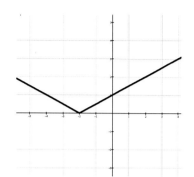

25. $f(x) = \begin{cases} -2x+1 & \text{for} \quad x \leq -2 \\ 5 & \text{for} \quad -2 < x < 3 \\ 2x-1 & \text{for} \quad x \geq 3 \end{cases}$

26. $f(x) = \begin{cases} -4x-2 & \text{for} \quad x \leq -6 \\ -2x+10 & \text{for} \quad -6 < x < \dfrac{4}{3} \\ 4x+2 & \text{for} \quad x \geq \dfrac{4}{3} \end{cases}$

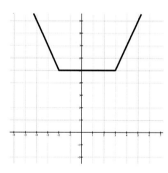
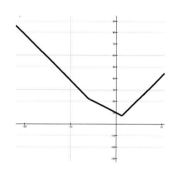

27. $f(x) = \begin{cases} -3x-4 & \text{for} \quad x \leq -\dfrac{1}{5} \\ 7x-2 & \text{for} \quad -\dfrac{1}{5} < x < \dfrac{3}{2} \\ 3x+4 & \text{for} \quad x \geq \dfrac{3}{2} \end{cases}$

28. $f(x) = \begin{cases} -6 & \text{for} \quad x \leq -1 \\ 2x-4 & \text{for} \quad -1 < x < 5 \\ 6 & \text{for} \quad x \geq 5 \end{cases}$

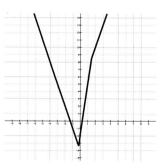
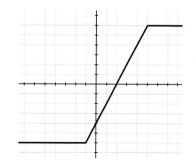

Lesson 2 — Synthetic Division

In an algebra class, you learned how to divide polynomials. Here is a reminder of what you should know. To become more familiar with this process, check out the videos online for MAT 1033.

Long Division of Polynomials

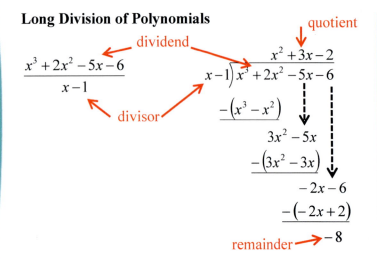

When one polynomial is divided by another, the result is a quotient and a remainder. If the remainder is 0, then the divisor is a factor of the dividend. The result can be written in two ways.

$$\frac{x^3+2x^2-5x-6}{x-1} = x^2+3x-2-\frac{8}{x-1}$$

and

$$x^3+2x^2-5x-6 = (x-1)(x^2+3x-2)-8$$

To streamline the process of long division, we use a procedure called **synthetic division**. Synthetic division is more efficient and condenses the long division process.

Step 1: Take the divisor, $x-1$, and solve $x-1=0$. $x=1$ is the number in the box. Write only the coefficients of the dividend x^3+2x^2-5x-6 to the right of the box. Variables are not written in the synthetic division procedure until the final answer.

$$\underline{1|}\quad 1\quad 2\quad -5\quad -6$$

Step 2: Start synthetic division by bringing down the leading coefficient.

$$\begin{array}{r|rrrr} 1 & 1 & 2 & -5 & -6 \\ & \downarrow & & & \\ \hline & 1 & & & \end{array}$$
Bring down 1.

Step 3: Multiply the leading coefficient by the number in the box. Now, add 2 + 1 to get 3. Multiply 3 by the number in the box. Now, add –5 and 3 to get –2. Multiply –2 by the number in the box. Finally, add –6 and –2 to get the last number –8.

$$\begin{array}{r|rrrr} 1 & 1 & 2 & -5 & -6 \\ & & 1 & 3 & -2 \\ \hline & 1 & 3 & -2 & -8 \end{array}$$

Step 4: The last row in synthetic division matches the coefficients of the quotient in the long division process, with the remainder –8 as the last number in that row. This is not a coincidence! It happens every time, therefore making the division process more efficient and concise. How do you find the quotient and remainder using the final row of synthetic division? The degree of the quotient will be one less than the degree of the dividend. Since we started with degree 3 in the dividend and divided by degree 1 in the divisor, the quotient will start with x^2. Therefore the quotient is x^2+3x-2. Write your answer in the following two ways.

$$\frac{dividend}{divisor} = quotient + \frac{remainder}{divisor} \quad \text{and} \quad dividend = (divisor)(quotient) + remainder$$

Synthetic division doesn't replace long division completely. In synthetic division, the divisor must be linear. If you want to divide $\frac{x^3 + 2x^2 - 5x - 6}{x^2 - 1}$, where the divisor is not linear, long division must be used.

When using synthetic division, it is important to write a 0 for a missing term. In the next example, be sure to put a 0 in the place of the missing x^3 term.

Example 1: Use synthetic division to divide the polynomials.

a. $(x^4 - 10x^2 - 2x + 4) \div (x + 4)$ *(dividend ÷ divisor)*

b. $(4x^3 + 10x^2 - 3x - 8) \div \left(x - \frac{1}{2}\right)$

Solution:

a.
```
-4 | 1    0   -10   -2    4
   |     -4    16  -24  104
   ----------------------------
     1   -4    6   -26  (108)  remainder
              ↑
           quotient
```

$x^4 - 10x^2 - 2x + 4 = (x + 4)(x^3 - 4x^2 + 6x - 26) + 108$
dividend = divisor · quotient + remainder

$$\frac{x^4 - 10x^2 - 2x + 4}{x + 4} = x^3 - 4x^2 + 6x - 26 + \frac{108}{x + 4}$$

dividend/divisor = quotient + remainder/divisor

b.
```
 1/2 | 4   10   -3   -8
     |      2    6   3/2
     --------------------
       4   12    3  -13/2
```

$4x^3 + 10x^2 - 3x - 8 = \left(x - \frac{1}{2}\right)(4x^2 + 12x + 3) - \frac{13}{2}$

$$\frac{4x^3 + 10x^2 - 3x - 8}{x - \frac{1}{2}} = 4x^2 + 12x + 3 - \frac{\frac{13}{2}}{x - \frac{1}{2}}$$

$$= 4x^2 + 12x + 3 - \frac{13}{2x - 1}$$

Remainder Theorem

If a polynomial $P(x)$ is divided by $x - c$, then the remainder r is equal to $P(c)$.

Proof of the Remainder Theorem:

Remember that $P(x) = (x-c)q(x) + r$ from polynomial division. Evaluate the polynomial for the value c. In other words, substitute c for x.

$$P(c) = (c-c)q(c) + r$$
$$= 0 \cdot q(c) + r$$
$$P(c) = r$$

Since $y = P(x)$, the remainder is the y-value on the graph of the polynomial at $x = c$.

From Example 1a, $P(x) = x^4 - 10x^2 - 2x + 4$. Because the remainder is $r = 108$, we can conclude from the Remainder Theorem $P(-4) = 108$. So if we were graphing this polynomial function, we would plot the point $(-4, 108)$.

Example 2: Use the Remainder Theorem to evaluate the following functions.

a. $P(x) = x^4 - 10x^2 - 2x + 4$; $P(2)$

b. $f(x) = 4x^3 + 10x^2 - 3x - 8$; $f(-1)$

Solution:

a.
```
2│  1   0   -10   -2    4
       2    4   -12  -28
    1   2   -6   -14  -24
```

$P(2) = -24$

b.
```
-1│  4   10   -3   -8
        -4   -6    9
     4    6   -9    1
```

$f(-1) = 1$

If $P(c) = 0$, then the y-value is zero, and the point is plotted on the x-axis at $(c, 0)$. This brings us to another important theorem regarding polynomials.

Factor Theorem

A polynomial $P(x)$ has a factor $x - c$ if and only if $P(c) = 0$.

If $r = 0$, then $x - c$ is a factor of $P(x)$.

Proof of the Factor Theorem:

If $r = 0$, then $\begin{array}{l} P(x) = (x-c)q(x) + 0 \\ P(x) = (x-c)q(x) \end{array}$. Therefore, $(x - c)$ is a factor of $P(x)$.

Conversely, if $(x-c)$ is a factor of $P(x)$, then division of $P(x)$ by $(x-c)$ yields a remainder of 0.

Example 3: Determine if the factors given are factors of the polynomials.

a. $P(x) = 2x^3 - 3x^2 + x + 6$; $(x+1)$ and $(x-3)$

b. $f(x) = x^4 - 10x^2 + 9$; $(x-2)$ and $(x+3)$

Solution:

a.
```
-1 | 2  -3   1   6          3 | 2  -3   1   6
   |    -2   5  -6             |     6   9  30
     2  -5   6   0               2   3  10  36
```

Since $P(-1) = 0$, $(x+1)$ is a factor of the polynomial. Since $P(3) \neq 0$, $(x-3)$ is not a factor.

b.
```
2 | 1   0  -10    0    9        -3 | 1   0  -10   0   9
  |     2    4  -12  -24           |    -3    9   3  -9
    1   2   -6  -12  -15             1  -3   -1   3   0
```

Since $f(2) \neq 0$, $(x-2)$ is not a factor of the polynomial. Since $P(-3) = 0$, $(x+3)$ is a factor.

If $x-c$ is a factor of the polynomial, then c is called a **zero of the polynomial**. Notice the location of the point $(c,0)$. Each zero of the polynomial is an x-intercept on the graph of the polynomial.

Example 4: Use the Factor Theorem to determine if the given values are zeros of $P(x)$.

a. $P(x) = 6x^3 - 4x^2 + 3x - 2$; $x = -3$ and $x = \dfrac{2}{3}$

b. $f(x) = x^3 + 2x^2 - 3x - 6$; $x = \sqrt{3}$ and $x = -2$

Solution:

a.
```
-3 | 6   -4    3    -2          2/3 | 6  -4   3  -2
   |     -18  66  -207              |     4   0   2
     6  -22   69  -209                 6   0   3   0
```

Since $P(-3) \neq 0$, $x = -3$ is not a zero of the polynomial. Since $P\left(\dfrac{2}{3}\right) = 0$, $x = \dfrac{2}{3}$ is a zero.

b.
$\sqrt{3}$	1	2	−3	−6
		$\sqrt{3}$	$3+2\sqrt{3}$	6
	1	$2+\sqrt{3}$	$2\sqrt{3}$	0

−2	1	2	−3	−6
		−2	0	6
	1	0	−3	0

Since $f(\sqrt{3})=0$ and $f(-2)=0$, both $x=\sqrt{3}$ and $x=-2$ are zeros of the polynomial.

Take a look at $f(x)=x^3+x+1$. By trial and error, can you find the zeros for the polynomial? No matter what integer value you choose, the remainder resulting from synthetic division is not zero. Does this mean the polynomial does not have a zero? Take a look at the graph. We see the graph crosses the x-axis, therefore the polynomial has a zero. To locate the real zero to any desired accuracy, we can use the Intermediate Value Theorem.

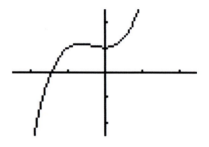

Intermediate Value Theorem

Let a and b be real numbers such that $a \neq b$. If $P(a)$ and $P(b)$ have opposite signs, then $P(x)$ has a real zero in the interval $[a,b]$.

Polynomial functions are continuous on the domain $(-\infty,\infty)$, which means there are no gaps or breaks in the graph. Evaluate the function for $x=-3$ and $x=0$. Since $f(-3)<0$ and $f(0)>0$, there is a real zero between -3 and 0. We can be more accurate by choosing $x=-2$ and $x=-1$. Since $f(-2)<0$ and $f(-1)>0$, there is a real zero between -2 and -1.

Example 5: Use the intermediate value theorem to verify the given polynomial has at least one zero in the specified interval.

a. $P(x)=x^3+x^2-6x$ $[-4,-2]$ b. $f(x)=3x^2-2x-11$ $[2,3]$

Solution:

a. Since $P(-4)<0$ and $P(-2)>0$, there is a real zero between -4 and -2.

b. Since $f(2)<0$ and $f(3)>0$, there is a real zero between 2 and 3.

Please note: If the sign of $f(a)$ and $f(b)$ are not opposite, the theorem does not apply. The answer would not include the statement, "There is no zero between $[a,b]$." The correct response would be the statement, "The Intermediate Value Theorem cannot be applied since $f(a)$ and $f(b)$ are not opposite signs."

In calculus, we will expand the idea of the Intermediate Value Theorem. If $f(x)$ is a continuous function on the interval $[a,b]$ and $f(a) \neq f(b)$, there exists a c in $[a,b]$ such that $f(c) = k$. The variable k is any y value between $f(a)$ and $f(b)$. If the graph of $f(x)$ can be drawn with no breaks or gaps, then $f(x)$ must take on every intermediate value from $f(a)$ and $f(b)$. This would include the value 0 if $f(a)$ and $f(b)$ are opposite signs.

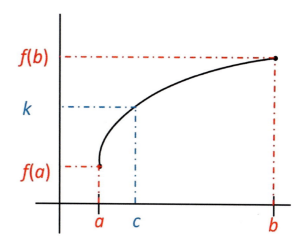

ALGEBRA REVIEW

1. Divide the polynomials using long division.

   a. $\dfrac{3x^4 + 2x^3 + x^2 + 4}{x^2 + 1}$

   b. $\dfrac{2x^5 + 3x^2 + 12}{x^3 - 3x - 4}$

   c. $\dfrac{-9x^6 + 7x^4 - 2x^3 + 5}{3x^2 - x + 1}$

2. Solve the quadratic equations.

   a. $x^2 - 6x - 16 = 0$

   b. $9x^2 = 24x + 16$

   c. $2x^2 = 6x$

   d. $3x^2 + 27 = 0$

   e. $(x-2)^2 = 25$

   f. $3x^2 - 12x + 13 = 0$

Lesson 2 Practice Exercises

In Exercises 1 – 12, divide using synthetic division. Write the answer as $P(x) = (x-c)q(x)+r$ and $\dfrac{P(x)}{x-c} = q(x) + \dfrac{r}{x-c}$.

1. $(2x^4 + 7x^3 - 6x^2 + x - 12) \div (x+3)$

2. $(3x^3 + x^2 + 4x - 1) \div (x+5)$

3. $(4x^5 + x^4 - 10x^3 + 13x^2 + 5x - 4) \div (x-2)$

4. $(-5x^4 - x^3 + 17x^2 + 3x - 20) \div (x-6)$

5. $(2x^3 - 3x^2 - 11x + 6) \div (x-3)$

6. $(15x^3 + 14x^2 - 3x - 2) \div (x+1)$

7. $(6x^4 + 10x^3 - 5x^2 - 3x + 1) \div \left(x + \dfrac{2}{3}\right)$

8. $(6x^3 - x^2 - 5x + 4) \div \left(x - \dfrac{2}{3}\right)$

9. $(5x^3 + 6x + 8) \div (x+2)$

10. $(x^3 + 7x^2 - 6) \div (x-4)$

11. $(x^3 - 64) \div (x-4)$

12. $(x^5 + 32) \div (x+2)$

In Exercises 13 – 18, use the remainder theorem to evaluate $P(x)$ for the given values.

13. $P(x) = 2x^4 + x^3 - x^2 + 3x + 3$; $P(-3)$ and $P(2)$

14. $P(x) = x^3 - 7x^2 + 5x - 6$; $P(-1)$ and $P(5)$

15. $P(x) = -2x^3 + 9x^2 - 11$; $P(-2)$ and $P\left(\dfrac{1}{2}\right)$

16. $P(x) = 4x^3 - 7x + 33$; $P\left(-\dfrac{1}{2}\right)$ and $P(3)$

17. $P(x) = x^5 - 10x^4 + 20x^3 - 5x - 100$; $P(-12)$ and $P(7)$

18. $P(x) = 5x^5 + 3x^3 - x + 8$; $P(-4)$ and $P(1)$

In Exercises 19 – 24, use the factor theorem to determine if the factor given is a factor of $P(x)$.

19. $(x+1)$ $P(x) = x^3 - 2x^2 - x + 2$

20. $(x-2)$ $P(x) = 2x^3 - 5x^2 + x + 2$

21. $(3x-1)$ $P(x) = 3x^3 + 8x^2 + 21x - 8$

22. $(5x+3)$ $P(x) = 5x^3 + 13x^2 - 9x - 9$

23. $(x+2)$ $P(x) = x^4 + 4x^3 + 2x^2 - 4x - 3$

24. $(x-4)$ $P(x) = x^4 - 26x^2 + 25$

In Exercises 25 – 30, use the factor theorem to determine if the given value is a zero of $P(x)$.

25. $x = 1$ $\quad P(x) = 2x^4 + 5x^3 + 3x^2 - 7x - 3$
26. $x = 3$ $\quad P(x) = 3x^3 + 5x^2 - 6x + 18$

27. $x = -2$ $\quad P(x) = x^5 + 2x^2 - 9x + 6$
28. $x = -\dfrac{1}{2}$ $\quad P(x) = 2x^3 + 3x^2 - 3x - 2$

29. $x = \dfrac{3}{2}$ $\quad P(x) = 2x^4 - x^3 + 3x^2 - 3x + 9$
30. $x = 2$ $\quad P(x) = x^3 - 7x + 6$

In Exercises 31 – 36, use the intermediate value theorem to verify the given polynomial has at least one zero in the specified interval.

31. $P(x) = 3x^3 - 8x^2 + x + 2$ $\quad [2,3]$
32. $P(x) = x^3 + 3x^2 - 9x - 13$ $\quad [-5,-4]$

33. $P(x) = 3x^3 - 10x + 9$ $\quad [-3,-2]$
34. $P(x) = 2x^5 - 7x + 1$ $\quad [1,2]$

35. $P(x) = 2x^4 - 4x^2 + 1$ $\quad [-1,0]$
36. $P(x) = x^3 - 5x^2 + 4$ $\quad [4,5]$

Solutions to Practice Exercises $\hfill$ **Lesson 2**

1. $2x^4 + 7x^3 - 6x^2 + x - 12 = (x+3)(2x^3 + x^2 - 9x + 28) - 96$

   $\dfrac{2x^4 + 7x^3 - 6x^2 + x - 12}{x+3} = 2x^3 + x^2 - 9x + 28 - \dfrac{96}{x+3}$

2. $3x^3 + x^2 + 4x - 1 = (x+5)(3x^2 - 14x + 74) - 371$

   $\dfrac{3x^3 + x^2 + 4x - 1}{x+5} = 3x^2 - 14x + 74 - \dfrac{371}{x+5}$

3. $4x^5 + x^4 - 10x^3 + 13x^2 + 5x - 4 = (x-2)(4x^4 + 9x^3 + 8x^2 + 29x + 63) + 122$

   $\dfrac{4x^5 + x^4 - 10x^3 + 13x^2 + 5x - 4}{x-2} = 4x^4 + 9x^3 + 8x^2 + 29x + 63 + \dfrac{122}{x-2}$

4. $-5x^4 - x^3 + 17x^2 + 3x - 20 = (x-6)(-5x^3 - 31x^2 - 169x - 1011) - 6086$

   $\dfrac{-5x^4 - x^3 + 17x^2 + 3x - 20}{x-6} = -5x^3 - 31x^2 - 169x - 1011 - \dfrac{6086}{x-6}$

5. $2x^3 - 3x^2 - 11x + 6 = (x-3)(2x^2 + 3x - 2) + 0$

   $\dfrac{2x^3 - 3x^2 - 11x + 6}{x-3} = 2x^2 + 3x - 2 + \dfrac{0}{x-3}$

6. $15x^3 + 14x^2 - 3x - 2 = (x+1)(15x^2 - x - 2) + 0$

   $\dfrac{15x^3 + 14x^2 - 3x - 2}{x+1} = 15x^2 - x - 2 + \dfrac{0}{x+1}$

7.
$$6x^4 + 10x^3 - 5x^2 - 3x + 1 = \left(x + \frac{2}{3}\right)\left(6x^3 + 6x^2 - 9x + 3\right) - 1$$

$$\frac{6x^4 + 10x^3 - 5x^2 - 3x + 1}{x + \frac{2}{3}} = 6x^3 + 6x^2 - 9x + 3 - \frac{1}{x + \frac{2}{3}}$$

$$\frac{6x^4 + 10x^3 - 5x^2 - 3x + 1}{x + \frac{2}{3}} = 6x^3 + 6x^2 - 9x + 3 - \frac{3}{3x + 2}$$

8.
$$6x^3 - x^2 - 5x + 4 = \left(x - \frac{2}{3}\right)\left(6x^2 + 3x - 3\right) + 2$$

$$\frac{6x^3 - x^2 - 5x + 4}{x - \frac{2}{3}} = 6x^2 + 3x - 3 + \frac{2}{x - \frac{2}{3}}$$

$$\frac{6x^3 - x^2 - 5x + 4}{x - \frac{2}{3}} = 6x^2 + 3x - 3 + \frac{6}{3x - 2}$$

9.
$$5x^3 + 6x + 8 = (x+2)\left(5x^2 - 10x + 26\right) - 44$$

$$\frac{5x^3 + 6x + 8}{x + 2} = 5x^2 - 10x + 26 - \frac{44}{x + 2}$$

10.
$$x^3 + 7x^2 - 6 = (x-4)\left(x^2 + 11x + 44\right) + 170$$

$$\frac{x^3 + 7x^2 - 6}{x - 4} = x^2 + 11x + 44 + \frac{170}{x - 4}$$

11.
$$x^3 - 64 = (x-4)\left(x^2 + 4x + 16\right) + 0$$

$$\frac{x^3 - 64}{x - 4} = x^2 + 4x + 16 + \frac{0}{x - 4}$$

12.
$$x^5 + 32 = (x+2)\left(x^4 - 2x^3 + 4x^2 - 8x + 16\right) + 0$$

$$\frac{x^5 + 32}{x + 2} = x^4 - 2x^3 + 4x^2 - 8x + 16 + \frac{0}{x + 2}$$

13. $P(-3) = 120$ and $P(2) = 45$

14. $P(-1) = -19$ and $P(5) = -31$

15. $P(-2) = 41$ and $P\left(\frac{1}{2}\right) = -9$

16. $P\left(-\frac{1}{2}\right) = 36$ and $P(3) = 120$

17. $P(-12) = -490792$ and $P(7) = -478$

18. $P(-4) = -5300$ and $P(1) = 15$

19. Yes   20. Yes   21. Yes   22. Yes   23. No   24. No

25. Yes   26. No   27. Yes   28. Yes   29. No   30. Yes

31. Verified   32. Verified   33. Verified   34. Verified   35. Verified   36. Verified

$P(2) = -4$	$P(-5) = -18$	$P(-3) = -42$	$P(1) = -4$	$P(-1) = -1$	$P(4) = -12$
$P(3) = 14$	$P(-4) = 7$	$P(-2) = 5$	$P(2) = 51$	$P(0) = 1$	$P(5) = 4$

Lesson 3 — Zeros of a Polynomial

If a polynomial has rational coefficients, then irrational zeros and complex zeros occur in pairs. These pairs are called conjugates. $a \pm b\sqrt{c}$ $a \pm bi$

Example 1: Given the following zeros, find a polynomial with rational coefficients.

$$-5,\ 4i,\ 1+\sqrt{3}$$

Solution: Only three zeros are given, but the irrational and complex zeros occur in pairs. Therefore, the polynomial has degree 5. Take each zero and write them as factors.

$$x=-5,\ \underbrace{x=4i,\ x=-4i}_{\text{conjugate pairs}},\ \underbrace{x=1+\sqrt{3},\ x=1-\sqrt{3}}_{\text{conjugate pairs}}$$

$$x+5 \quad x-4i \quad x+4i \quad x-1-\sqrt{3} \quad x-1+\sqrt{3}$$

Now, multiply these factors to find the polynomial. For easier multiplication, multiply the conjugate pairs first.

$$P(x) = (x+5)(x-4i)(x+4i)(x-1-\sqrt{3})(x-1+\sqrt{3})$$

$$P(x) = (x+5)(x^2 - 16i^2)(x^2 - 2x - 2) \qquad \text{***Remember: } i^2 = -1\text{***}$$

$$P(x) = (x+5)(x^2 + 16)(x^2 - 2x - 2)$$

$$P(x) = (x^3 + 5x^2 + 16x + 80)(x^2 - 2x - 2)$$

$$P(x) = x^5 + 3x^4 + 4x^3 + 38x^2 - 192x - 160$$

Sometimes the zeros of a polynomial are repeated as in the polynomial, $x^2 + 4x + 4 = (x+2)^2$. Therefore, $x = -2$ is a zero with multiplicity 2.

Multiplicity

If $(x-c)^k$ is a factor of a polynomial function, $x = c$ has **multiplicity** k.

Multiplicity is the number of occurrences of the zero in a function.

Example 2: Given the following zeros, find a polynomial with rational coefficients of degree 4.

$6, 5-\sqrt{2}$

Solution: Since only two zeros are given, multiplicity must be involved. The irrational zero has a conjugate pair, so $x = 6$ must have multiplicity 2 for the polynomial to result in degree 4. Start by writing each zero as a factor.

Zeros: $x=6$, $x=5-\sqrt{2}$, $x=5+\sqrt{2}$

Factors: $x-6$ $x-5+\sqrt{2}$ $x-5-\sqrt{2}$

Now, multiply these factors to find the polynomial. Don't forget to write the multiplicity of 2 for factor $(x-6)$.

$$P(x) = (x-6)^2(x-5+\sqrt{2})(x-5-\sqrt{2})$$

$$P(x) = (x^2-12x+36)(x^2-10x+23)$$

$$P(x) = x^4 - 22x^3 + 179x^2 - 636x + 828$$

What if a polynomial is given and we want to find all the zeros of the polynomial? We could start with algebra methods. Using the factoring steps from algebra will work great if we are given a quadratic polynomial, the difference of squares, the sum/difference of cubes. There are algebra formulas for those polynomials. But what if we are given a polynomial like the solution to example 2 and we need to work backwards to find the zeros?

From Lesson 2, the Factor Theorem stated that the remainder would be 0 if $x - c$ were a factor of the polynomial. All we have to do is perform synthetic division on all possible values and find the zeros. Since there are infinite values to choose from, we need a way to determine which values to try. Remember we have three different kinds of zeros to find: rational, irrational, and complex. The following theorem will help us find the rational zeros. Rational zeros are numbers that can be written as fractions.

Rational Zeros Theorem

Given a polynomial $P(x)$ with integer coefficients. The rational zeros, if they exist, of $P(x)$ must be of the form $\frac{p}{q}$, where p is a factor of the constant term and q is a factor of the leading coefficient. The rational number $\frac{p}{q}$ must be in lowest terms.

If the final quotient does not factor, then use the quadratic formula to find remaining zeros. These zeros could be irrational or complex.

$$x = \frac{-b \pm \sqrt{b^2 - 4ac}}{2a}$$ Quadratic Formula

Remember from algebra: Discriminant
If $b^2 - 4ac > 0$, the zeros are real. These real zeros are either rational or irrational.
If $b^2 - 4ac < 0$, the zeros are complex.

Example 3: Use the Rational Zeros Theorem to list the possible rational zeros for the following polynomials. Then factor the polynomial completely.

a. $P(x) = 3x^4 - 10x^3 + 15x^2 + 20x - 8$ **b.** $f(x) = x^3 - 4x^2 - 7x + 10$

Solution:

a. p must be a factor of -8 and q must be a factor of 3.

$$\frac{factors\ of\ -8}{factors\ of\ 3} = \frac{\pm 1, \pm 2, \pm 4, \pm 8}{\pm 1, \pm 3}$$

Written in lowest terms, $\frac{p}{q} = \pm 1, \pm \frac{1}{3}, \pm 2, \pm \frac{2}{3}, \pm 4, \pm \frac{4}{3}, \pm 8, \pm \frac{8}{3}$ is the list of possible rational zeros.

Now use synthetic division to determine which numbers from the list result in a zero remainder. There is no need to perform synthetic division for $x = 3$ or $x = \frac{1}{2}$. These values are not in the list of possible rational zeros.

```
-1 | 3  -10   15   20   -8
   |     -3   13  -28    8
   |_____
     3  -13   28   -8    0
```
quotient of the zero

When you find one zero remainder, do not return to the original polynomial to continue finding the other zeros. Use the "reduced" polynomial, which is the quotient of the zero.

```
          quotient of the zero
1/3 | 3  -13   28   -8
    |      1   -4    8
    |_____
      3  -12   24    0
```

Therefore, $x = -1$ and $x = \frac{1}{3}$ are rational zeros of the polynomial. The final quotient is quadratic, which is factored as $3(x^2 - 4x + 8)$. The factored form of the polynomial is $3(x+1)\left(x - \frac{1}{3}\right)(x^2 - 4x + 8)$ or $(x+1)(3x-1)(x^2 - 4x + 8)$. Notice, this polynomial has two rational zeros and two complex zeros. To find the complex zeros, use the quadratic formula.

$$\frac{-b \pm \sqrt{b^2 - 4ac}}{2a} = \frac{-(-4) \pm \sqrt{(-4)^2 - 4(1)(8)}}{2(1)} = \frac{4 \pm 4i}{2} = 2 \pm 2i$$ Notice the conjugate pairs.

To write the polynomial as linear factors, $P(x) = (x+1)(3x-1)(x - 2 - 2i)(x - 2 + 2i)$.

---

b. p must be a factor of 10 and q must be a factor of 1.

$$\frac{\text{factors of } 10}{\text{factors of } 1} = \frac{\pm 1, \pm 2, \pm 5, \pm 10}{\pm 1}$$

Written in lowest terms, $\frac{p}{q} = \pm 1, \pm 2, \pm 5, \pm 10$ is the list of possible rational zeros.

Notice that ± 1 will always be in the list of possible rational zeros. There is a quick way to check for $x = 1$. If the coefficients of the polynomial add to 0, then $x = 1$ will be a zero of the polynomial. For $f(x) = x^3 - 4x^2 - 7x + 10$, the coefficients add to 0. Therefore, we know that synthetic division with $x = 1$ will result in a zero remainder.

```
 1 | 1  -4  -7   10            -2 | 1  -3  -10        Final quotient yields x - 5.
   |     1  -3  -10                |    -2   10
   |_____                |_____
     1  -3  -10   0                  1  -5    0
```
quotient of the zero

Therefore, $x = 1$, $x = -2$, and $x = 5$ are rational zeros of the polynomial. The completely factored form is $f(x) = (x-1)(x+2)(x-5)$.

For a polynomial of degree n, the polynomial has at most n zeros. Those zeros can be combinations of rational, irrational, and complex. The zeros could be positive, negative, or 0. The following theorem gives the possible combinations for zeros of a given polynomial.

Descartes' Rule of Signs

Given a polynomial $P(x)$ written in descending or ascending order with real coefficients and a nonzero constant term, the number of positive zeros is the same as the number of variations of sign in $P(x)$, or an even number less. The number of negative zeros is the same as the number variations of sign in $P(-x)$, or an even number less.

Example 4: Use Descartes' rule of signs to count the number of possible positive, negative, and complex zeros of the polynomial.

a. $P(x) = 4x^4 + 4x^3 - 7x^2 - 4x + 3$
b. $f(x) = 2x^4 + x^3 - 4x^2 - 3x$

Solution:

a. $P(x) = 4x^4 + 4x^3 - 7x^2 - 4x + 3$ There are two sign changes in $P(x)$. According to

Descartes' Rule of Signs, there are 2 or 0 positive zeros.

$$P(-x) = 4(-x)^4 + 4(-x)^3 - 7(-x)^2 - 4(-x) + 3$$
$$= 4x^4 - 4x^3 - 7x^2 + 4x + 3$$

There are two sign changes in $P(-x)$. According to Descartes Rule of Signs, there are 2 or 0 negative zeros. To organize this information, use a table. Using 2 or 0 for both, we have four possible combinations.

Positive	Negative	Complex
2	2	0
2	0	2
0	2	2
0	0	4

Each row total will add to the degree of the polynomial. Since the degree is 4, each row should total 4. This helps us determine the possible number of complex zeros in the polynomial. Remember: complex zeros occur in pairs. Therefore, the possible number of complex zeros will never be odd!

b. $f(x) = 2x^4 + x^3 - 4x^2 - 3x$ has a zero constant term. Factor out the GCF (greatest common factor), then use Descartes' Rule of Signs. $f(x) = x(2x^3 + x^2 - 4x - 3)$. We can see from the GCF that $x = 0$ is a factor with multiplicity of 1. Now, analyze $2x^3 + x^2 - 4x - 3$ with Descartes Rule of Signs.

There is one sign change, therefore there 1 positive zero.

$$f(-x) = 2(-x)^3 + (-x)^2 - 4(-x) - 3$$
$$= -2x^3 + x^2 + 4x - 3$$

Since there are two sign changes, there are 2 or 0 negative zeros.

$x = 0$	Positive	Negative	Complex
1	1	2	0
1	1	0	2

Each row will add to degree $n = 4$.

The first column represents the GCF of x that has multiplicity of 1.

Example 5: Solve the equation.

$3x^4 - 5x^3 - 9x^2 + 9x + 10 = 0$

Solution: Solving an equation that is equal to 0 means you are actually finding all the zeros of the polynomial. Use the theorems to determine the zeros. These zeros will solve the equation.

Let $P(x) = 3x^4 - 5x^3 - 9x^2 + 9x + 10$.

According to the rational zeros theorem, $\dfrac{p}{q} = \dfrac{\pm 1, \pm 2, \pm 5, \pm 10}{\pm 1, \pm 3} \pm 1, \pm \dfrac{1}{3}, \pm 2, \pm \dfrac{2}{3}, \pm 5, \pm \dfrac{5}{3}, \pm 10, \pm \dfrac{10}{3}$

$P(x) = 3x^4 - 5x^3 - 9x^2 + 9x + 10$ has two sign changes. $P(-x) = 3x^4 + 5x^3 - 9x^2 - 9x + 10$ also has two sign changes. According to Descartes Rule of Signs, this polynomial has 2 or 0 positive zeros and 2 or 0 negative zeros.

Use the Factor Theorem to determine which numbers from the list are actual zeros of the polynomial.

The coefficients of $P(x)$ do not add to zero, therefore $x = 1$ is not a zero of the polynomial. If you add the coefficients of $P(-x) = 3x^4 + 5x^3 - 9x^2 - 9x + 10$, you get zero. Therefore, $x = -1$ is a zero of the polynomial. Start synthetic division with $x = -1$.

```
-1 | 3  -5  -9   9   10              2 | 3  -8  -1   10            Final quotient yields
   |    -3   8   1  -10                |     6  -4  -10             3x² - 2x - 5
   |  3  -8  -1  10    0                |  3  -2  -5    0           (3x - 5)(x + 1)
```

The factored form is $(x+1)^2(x-2)(3x-5)$. To solve the equation, set each factor equal to zero and solve for x. The solution set is $\left\{ -1, \dfrac{5}{3}, 2 \right\}$.

Remember: If the final quotient does not factor, use the quadratic formula to solve the equation. The solutions could be irrational or complex, which always occur in conjugate pairs.

Example 6: Solve the equation.

$x^4 + x^3 - x - 1 = 0$

Solution: Find all the zeros of the polynomial. Starting with synthetic division, $x = 1$ will have a remainder of 0 since the sum of the coefficients equal 0.

```
1 | 1  1   0  -1  -1              -1 | 1  2   2   1
  |    1   2   2   1                 |    -1  -1  -1
  | 1  2   2   1   0                  | 1  1   1   0
```

The final quotient yields $x^2 + x + 1$, which does not factor. Using the quadratic formula, we can find two more solutions. The solution set for the equation is $\left\{ -1, 1, \dfrac{-1 \pm i\sqrt{3}}{2} \right\}$.

Lesson 3 — Practice Exercises

In Exercises 1 – 14, find a polynomial having rational coefficients with the given degree and zeros.

1. Degree 2, $x = 5, x = -6$
2. Degree 3, $x = 1, x = 3, x = -7$
3. Degree 3, $x = -4, x = 1 + 5i$
4. Degree 3, $x = 2, x = 7 - 3i$
5. Degree 3, $x = \sqrt{2}, x = 9$
6. Degree 3, $x = -\sqrt{5}, x = 1$
7. Degree 4, $x = 2i, x = \sqrt{3}$
8. Degree 4, $x = -6i, x = \sqrt{2}$
9. Degree 4, $x = -1, x = \sqrt{6}$
10. Degree 4, $x = 9, x = 4i$
11. Degree 4, $x = -10, x = -i$
12. Degree 4, $x = -1, x = 2 - \sqrt{3}$

In Exercises 13 – 30, determine the following.

a. Use the rational zeros theorem to list all possible rational zeros of the polynomial.
b. Use Descartes' rule of signs to count the number of possible positive, negative, and complex zeros of the polynomial.
c. Find all the zeros of the polynomial.
d. Factor the polynomial completely.

13. $P(x) = x^4 - 5x^3 + 20x - 16$
14. $P(x) = x^3 - 13x + 12$
15. $P(x) = 4x^3 + 8x^2 - 3x - 9$
16. $P(x) = 2x^3 - 3x^2 - 9x + 10$
17. $P(x) = 3x^4 - 4x^3 + x^2 + 6x - 2$
18. $P(x) = 5x^4 + 4x^3 - 5x - 4$
19. $P(x) = 2x^3 - 3x^2 - 11x + 6$
20. $P(x) = 15x^3 + 14x^2 - 3x - 2$
21. $P(x) = 6x^4 - 7x^3 - 73x^2 + 14x + 24$
22. $P(x) = 2x^4 - 5x^3 - 5x^2 + 5x + 3$
23. $P(x) = 3x^3 + 7x^2 - 22x - 8$
24. $P(x) = 24x^3 - 26x^2 + 9x - 1$
25. $P(x) = 12x^3 + 16x^2 - 5x - 3$
26. $P(x) = 3x^4 - 11x^3 + 10x - 4$
27. $P(x) = 4x^4 + 40x^3 - 93x^2 + 30x - 72$
28. $P(x) = 4x^4 - 42x^3 - 70x^2 - 21x - 36$

In Exercises 29 – 32, solve the equations.

29. $x^3 + 12x^2 + 21x + 10 = 0$
30. $6x^3 + 25x^2 - 24x + 5 = 0$
31. $3x^3 - 8x^2 = 8x - 8$
32. $2x^4 + 3x^3 + 15 = 11x^2 + 9x$

Solutions to Practice Exercises **Lesson 3**

1. $x^2 + x - 30$
2. $x^3 + 3x^2 - 25x + 21$
3. $x^3 + 2x^2 + 18x + 104$
4. $x^3 - 16x^2 + 86x - 116$
5. $x^3 - 9x^2 - 2x + 18$
6. $x^3 - x^2 - 5x + 5$
7. $x^4 + x^2 - 12$
8. $x^4 + 34x^2 - 72$
9. $x^4 + 2x^3 - 5x^2 - 12x - 6$
10. $x^4 - 18x^3 + 97x^2 - 288x + 1296$
11. $x^4 + 20x^3 + 101x^2 + 20x + 100$
12. $x^4 - 2x^3 - 6x^2 - 2x + 1$

13. a. $\pm 1, \pm 2, \pm 4, \pm 8, \pm 16$
    b.
P	N	C
3	1	0
1	1	2

    c. $\{-2, 1, 2, 4\}$
    d. $(x+2)(x-1)(x-2)(x-4)$

14. a. $\pm 1, \pm 2, \pm 3, \pm 4, \pm 6, \pm 12$
    b.
P	N	C
2	1	0
0	1	2

    c. $\{-4, 1, 3\}$
    d. $(x+4)(x-1)(x-3)$

15. a. $\pm 1, \pm \dfrac{1}{2}, \pm \dfrac{1}{4}, \pm 3, \pm \dfrac{3}{2}, \pm \dfrac{3}{4}, \pm 9, \pm \dfrac{9}{2}, \pm \dfrac{9}{4}$
    b.
P	N	C
1	2	0
1	0	2

    c. $\left\{-\dfrac{3}{2}, 1\right\}$
    d. $(2x+3)^2(x-1)$ or $4\left(x+\dfrac{3}{2}\right)^2(x-1)$

16. a. $\pm 1, \pm \dfrac{1}{2}, \pm 2, \pm 5, \pm \dfrac{5}{2}, \pm 10$
    b.
P	N	C
2	1	0
0	1	2

    c. $\left\{-2, 1, \dfrac{5}{2}\right\}$
    d. $(x+2)(x-1)(2x-5)$ or $2(x+2)(x-1)\left(x-\dfrac{5}{2}\right)$

17. a. $\pm 1, \pm \dfrac{1}{3}, \pm 2, \pm \dfrac{2}{3}$
    b.
P	N	C
3	1	0
1	1	2

    c. $\left\{-1, \dfrac{1}{3}, 1 \pm i\right\}$
    d. $(x+1)(3x-1)(x^2 - 2x + 2)$ or $3(x+1)\left(x - \dfrac{1}{3}\right)(x^2 - 2x + 2)$

18. a. $\pm 1, \pm \frac{1}{5}, \pm 2, \pm \frac{2}{5}, \pm 4, \pm \frac{4}{5}$
b.
P	N	C
1	3	0
1	1	2

c. $\left\{-\frac{4}{5}, 1, \frac{-1 \pm i\sqrt{3}}{2}\right\}$

d. $(5x+4)(x-1)(x^2+x+1)$ or $5\left(x+\frac{4}{5}\right)(x-1)(x^2+x+1)$

19. a. $\pm 1, \pm \frac{1}{2}, \pm 2, \pm 3, \pm \frac{3}{2}, \pm 6$
b.
P	N	C
2	1	0
0	1	2

c. $\left\{-2, \frac{1}{2}, 3\right\}$

d. $(x+2)(2x-1)(x-3)$ or $2(x+2)\left(x-\frac{1}{2}\right)(x-3)$

20. a. $\pm 1, \pm \frac{1}{3}, \pm \frac{1}{5}, \pm \frac{1}{15}, \pm 2, \pm \frac{2}{3}, \pm \frac{2}{5}, \pm \frac{2}{15}$
b.
P	N	C
1	2	0
1	0	2

c. $\left\{-1, -\frac{1}{3}, \frac{2}{5}\right\}$

d. $(x+1)(3x+1)(5x-2)$ or $15(x+1)\left(x+\frac{1}{3}\right)\left(x-\frac{2}{5}\right)$

21. a. $\pm 1, \pm \frac{1}{2}, \pm \frac{1}{3}, \pm \frac{1}{6}, \pm 2, \pm \frac{2}{3}, \pm 3, \pm \frac{3}{2}, \pm 4, \pm \frac{4}{3}, \pm 6, \pm 8, \pm \frac{8}{3}, \pm 12, \pm 24$
b.
P	N	C
2	2	0
0	2	2
2	0	2
0	0	4

c. $\left\{-3, -\frac{1}{2}, \frac{2}{3}, 4\right\}$

d. $(x+3)(2x+1)(3x-2)(x-4)$ or $6(x+3)\left(x+\frac{1}{2}\right)\left(x-\frac{2}{3}\right)(x-4)$

22. a. $\pm 1, \pm \frac{1}{2}, \pm 3, \pm \frac{3}{2}$
b.
P	N	C
2	2	0
0	2	2
2	0	2
0	0	4

c. $\left\{-1, -\frac{1}{2}, 1, 3\right\}$

d. $(x+1)(2x+1)(x-1)(x-3)$ or $2(x+1)\left(x+\frac{1}{2}\right)(x-1)(x-3)$

23. a. $\pm 1, \pm \frac{1}{3}, \pm 2, \pm \frac{2}{3}, \pm 4, \pm \frac{4}{3}, \pm 8, \pm \frac{8}{3}$
b.
P	N	C
1	2	0
1	0	2

c. $\left\{-4, -\frac{1}{3}, 2\right\}$

d. $(x+4)(3x+1)(x-2)$ or $3(x+4)\left(x+\frac{1}{3}\right)(x-2)$

24 a. $\pm 1, \pm \frac{1}{2}, \pm \frac{1}{3}, \pm \frac{1}{4},$
$\pm \frac{1}{6}, \pm \frac{1}{8}, \pm \frac{1}{12}, \pm \frac{1}{24}$

b.
P	N	C
3	0	0
1	0	2

c. $\left\{\frac{1}{4}, \frac{1}{3}, \frac{1}{2}\right\}$

d. $(4x-1)(3x-1)(2x-1)$ or $24\left(x-\frac{1}{4}\right)\left(x-\frac{1}{3}\right)\left(x-\frac{1}{2}\right)$

25 a. $\pm 1, \pm \frac{1}{2}, \pm \frac{1}{3}, \pm \frac{1}{4}, \pm \frac{1}{6},$
$\pm \frac{1}{12}, \pm 3, \pm \frac{3}{2}, \pm \frac{3}{4}$

b.
P	N	C
1	2	0
1	0	2

c. $\left\{-\frac{3}{2}, -\frac{1}{3}, \frac{1}{2}\right\}$

d. $(2x+3)(3x+1)(2x-1)$ or $12\left(x+\frac{3}{2}\right)\left(x+\frac{1}{3}\right)\left(x-\frac{1}{2}\right)$

26 a. $\pm 1, \pm \frac{1}{3}, \pm 2, \pm \frac{2}{3},$
$\pm 4, \pm \frac{4}{3}$

b.
P	N	C
3	1	0
1	1	2

c. $\left\{-1, \frac{2}{3}, 2 \pm \sqrt{2}\right\}$

d. $(x+1)(3x-2)(x^2-4x+2)$ or $3(x+1)\left(x-\frac{2}{3}\right)(x^2-4x+2)$

27 a. $\pm 1, \pm \frac{1}{2}, \pm \frac{1}{4}, \pm 2, \pm 3,$
$\pm \frac{3}{2}, \pm \frac{3}{4}, \pm 4, \pm 6, \pm 8,$
$\pm 9, \pm \frac{9}{2}, \pm \frac{9}{4}, \pm 12, \pm 18,$
$\pm 24, \pm 36, \pm 72$

b.
P	N	C
3	1	0
1	1	2

c. $\left\{-12, 2, \pm \frac{i\sqrt{3}}{2}\right\}$

d. $(x+12)(x-2)(4x^2+3)$

28 a. $\pm 1, \pm \frac{1}{2}, \pm \frac{1}{4}, \pm 2, \pm 3, \pm \frac{3}{2},$
$\pm \frac{3}{4}, \pm 4, \pm 6, \pm 9, \pm \frac{9}{2}, \pm \frac{9}{4},$
$\pm 12, \pm 18, \pm 36$

b.
P	N	C
1	3	0
1	1	2

c. $\left\{-\frac{3}{2}, 12, \pm \frac{i\sqrt{2}}{2}\right\}$

d. $(2x+3)(x-12)(2x^2+1)$

29. $\{-10, -1\}$ 30. $\left\{-5, \frac{1}{3}, \frac{1}{2}\right\}$ 31. $\left\{\frac{2}{3}, 1-\sqrt{5}, 1+\sqrt{5}\right\}$ 32. $\left\{-\frac{5}{2}, -\sqrt{3}, 1, \sqrt{3}\right\}$

Lesson 4 — Graphing a Polynomial

What is a polynomial? We can find zeros of a polynomial. We can find a polynomial if we know the zeros. But can you define a polynomial?

> A **polynomial** is a function written in the form $P(x) = a_n x^n + a_{n-1} x^{n-1} + \ldots + a_2 x^2 + a_1 x + a_0$
>
> The coefficients are real numbers and the exponents are whole numbers. The number a_n is the leading coefficient, and the number a_0 is the constant term. The degree of a polynomial is n, the largest exponent.

$f(x) = 3^x$, $f(x) = |x+5|$, and $f(x) = 3x^{-2}$ are not polynomials.

Polynomials are predictable. You can determine the number of zeros and the end behavior of the graph just by looking at the polynomial function. The graph of a polynomial is smooth and continuous. Smooth describes the "turning points" of the curve. These turning points have no sharp turns or jagged edges. Continuous is a word that will be defined in calculus. For now, we will use the word continuous to describe the graph can be drawn without picking up your pencil. In other words, a polynomial is continuous because there are no gaps or breaks in the graph. The domain includes all real numbers, $(-\infty, \infty)$.

There are at most n zeros in a polynomial. For a polynomial with degree 4, there are at most 4 zeros. Remember: Zeros of a polynomial are x–intercepts. Therefore, a polynomial has at most n x–intercepts.

Leading Term Test

Given a polynomial $P(x) = a_n x^n + a_{n-1} x^{n-1} + \ldots + a_2 x^2 + a_1 x + a_0$ with leading term $a_n x^n$. If $n \geq 1$, then the following can be used to determine the behavior of the graph as $x \to \infty$ and $x \to -\infty$.

1.) If n is even, the ends of the graph will point in the same direction.

$a_n > 0$ Leading Term is Positive

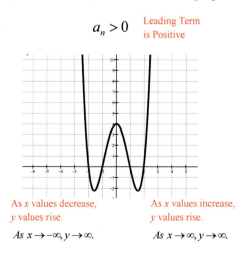

As x values decrease, y values rise.
As $x \to -\infty, y \to \infty$.

As x values increase, y values rise.
As $x \to \infty, y \to \infty$.

$a_n < 0$ Leading Term is Negative

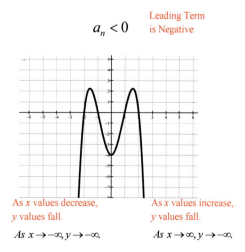

As x values decrease, y values fall.
As $x \to -\infty, y \to -\infty$.

As x values increase, y values fall.
As $x \to \infty, y \to -\infty$.

2.) If *n* is odd, the ends of the graph will point in the opposite direction.

$a_n > 0$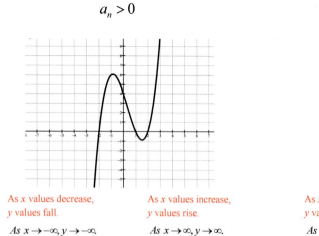

$a_n < 0$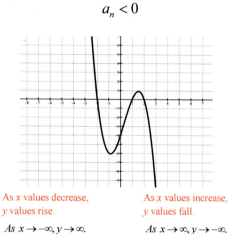

As *x* values decrease, *y* values fall.
As $x \to -\infty, y \to -\infty$.

As *x* values increase, *y* values rise.
As $x \to \infty, y \to \infty$.

As *x* values decrease, *y* values rise.
As $x \to -\infty, y \to \infty$.

As *x* values increase, *y* values fall.
As $x \to \infty, y \to -\infty$.

A polynomial has at most $n-1$ turning points, also called **relative extrema**. We can find the turning points using the graphing calculator. After you enter your polynomial, press **2ND** and **TRACE**.

Under the CALCULATE menu, you will find 3: minimum and 4: maximum. You will learn how to find relative extrema without a calculator in calculus.

> **Example 1:** Find the following information for the polynomial $p(x) = 2x^4 + x^3 - 6x^2 + x + 2$.

1. Determine the maximum number of zeros.

2. Determine the end behavior of the graph using the Leading Term Test.

3. Find the turning points of the graph.

Solution: The given polynomial has degree $n = 4$. Therefore, this polynomial has at most 4 zeros. Since the degree is even, the ends of the graph will point in the same direction. To determine which direction, we look at the leading coefficient. Since $a_n = 2 > 0$, both ends of the graph will point up. In other words, *as* $x \to \infty, y \to \infty$ and *as* $x \to -\infty, y \to \infty$.

Now, enter the polynomial into the graphing calculator. $y = 2x^4 + x^3 - 6x^2 + x + 2$
If you look at the graph, you will see 3 turning points, 2 relative minimums and 1 relative maximum. Find the CALCULATE menu and press #3. Your calculator requires you to move the cursor to the left of the relative minimum. Then press enter. Then move the cursor by pressing the arrow keys to the right of the relative minimum. Then press enter. Press enter one more time, and boom! The first relative minimum is $(-1.46, -6.28)$. The second relative minimum $(1, 0)$ can be found using the same steps.

Now find the relative maximum. Find the CALCULATE menu and press #4. Your calculator requires you to move the cursor to the left of the relative maximum. Then press enter. Then move the cursor by pressing the arrow keys to the right of the relative maximum. Then press enter. Press enter one more time, and the relative maximum is $(0.09, 2.04)$. Write the relative maxima and minima as turning points.

From this example, we have a really good idea of how to graph $p(x) = 2x^4 + x^3 - 6x^2 + x + 2$. Now, we just need a few details to make it complete. In fact, you can follow the 5 guidelines when graphing a polynomial.

Guidelines for Graphing Polynomials

1. Use the Leading Term Test to determine the end behavior of the graph.

2. Find the zeros of the polynomial and plot the x-intercepts.

3. Find the turning points, using the relative minimum and relative maximum feature in your calculator.

4. Find $p(0)$ and plot the y-intercept.

5. Plot additional points if necessary to finish the smooth continuous curve.

Example 2: Graph the polynomial $p(x) = 2x^4 + x^3 - 6x^2 + x + 2$.

Solution: Using the information found in Example 1, we know the end behavior and the location of the turning points. Now, we need to use the theorems in the previous lessons to determine the zeros. Start with the Rational Zeros Theorem to list possible rational zeros.

p must be a factor of 2 and q must be a factor of 2.

$$\frac{\text{factors of } 2}{\text{factors of } 2} = \frac{\pm 1, \pm 2}{\pm 1, \pm 2}$$

Written in lowest terms, $\frac{p}{q} = \pm 1, \pm \frac{1}{2}, \pm 2$ is the list of possible rational zeros.

Now use synthetic division to determine which numbers from the list result in a zero remainder. Notice, the coefficients add to 0. Therefore, $x = 1$ will result in a zero remainder.

```
1│ 2   1  -6   1   2          -2│ 2   3  -3  -2              Final quotient yields
       2   3  -3  -2                 -4   2   2              $2x^2 - x - 1$
    ─────────────────            ─────────────────
    2   3  -3  -2   0            2  -1  -1   0               $(2x+1)(x-1)$
```

Therefore, $x = -2$, $x = -\frac{1}{2}$, and $x = 1$ are rational zeros of the polynomial. We need to plot the zeros on the x-axis. $(-2, 0)$ $\left(-\frac{1}{2}, 0\right)$ $(1, 0)$

Using the Remainder Theorem, we can find other points to make the graph complete. We want a smooth and continuous curve. Choose any x value between the zeros and evaluate for the polynomial.

$p(-3) = 80$ $p(-1) = -4$ $p(0) = 2$ $p(2) = 20$

Notice points $(-3, 80)$ and $(2, 20)$ are extreme values that are not plotted on this scale. These points show the end behavior is very steep. Take a look at the point $(1, 0)$, which is an x–intercept since $x = 1$ is a zero of the polynomial. This point happens to be a relative minimum also. This brings us to a special property of multiplicity. In Lesson 3, we defined the word multiplicity as the number of occurrences of the zero in a function. Here, we will see the effects of multiplicity on the graph. If you look at the factored form of the polynomial, you can see that $x = 1$ has a multiplicity of 2. This means the factor $(x - 1)$ is repeated.

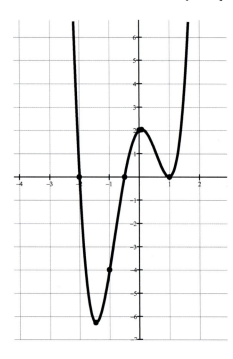

$$p(x) = 2x^4 + x^3 - 6x^2 + x + 2$$
$$= (x+2)(2x+1)(x-1)^2$$

Both $x = -2$ and $x = -\frac{1}{2}$ have multiplicity of 1. How does the multiplicity affect the graph at these x–intercepts?

Multiplicity of Zeros

Given a polynomial function $p(x)$ with factors of the form $(x - c)^k$, where c is a real number:

1. If k is odd, the graph will cross through the x-axis at $(c, 0)$.

2. If k is even, the graph will have a turning point at $(c, 0)$.

Zeros	Multiplicity	Cross/Turning Point
$x = 1$	1	Cross
$x = -\dfrac{1}{2}$	1	Cross
$x = 1$	2	Turning Point

Example 3: Find the zeros of the polynomial function and state the multiplicity of each. Then determine the behavior of the graph at each zero.

$$f(x) = (x+3)^3(3x-2)^4(x+5)^2$$

Solution: Take each factor and set it equal to zero. Solve for x.

$x + 3 = 0 \qquad 3x - 2 = 0 \qquad x + 5 = 0$

$x = -3 \qquad\quad x = \dfrac{2}{3} \qquad\quad x = -5$

Zeros	Multiplicity	Cross/Turning Point
$x = -3$	3	Cross
$x = \dfrac{2}{3}$	4	Turning Point
$x = -5$	2	Turning Point

In this example, the graph will "flatten" near the zero $x = \dfrac{2}{3}$ because of the higher multiplicity. In general, for larger multiplicity the graph will flatten out at the zero.

Can you determine the end behavior of the polynomial in Example 3? First, we need the degree and the leading coefficient. Add the multiplicities to find the degree is $n = 9$. Since n is odd, the ends of the graph will point in the opposite direction. To determine the leading coefficient we need to multiply all the factors. This would be a long and tedious process, and it really isn't necessary. We don't need the actual number coefficient. We only need to know if it is positive or negative. Take a look at the factors. If we were to multiply them out completely, a negative x would never occur. None of the factors include $-x$. Therefore, we can determine $a_n > 0$. With an odd degree and positive leading coefficient, we can conclude the graph will go up on the far right and down on the far left.

As $x \to \infty, y \to \infty$, and as $x \to -\infty, y \to -\infty$.

Example 4: Follow the 5 guidelines to graph the polynomial.

$f(x) = -x^3 + 2x^2 + 4x - 8$

Solution:

1. Use the Leading Term Test to determine the end behavior of the graph.

   $n = 3 \rightarrow odd$ and $a_n = -1 < 0$

   as $x \rightarrow \infty, y \rightarrow -\infty$
   as $x \rightarrow -\infty, y \rightarrow \infty$ Far left on the graph rises (∞) and far right on the graph falls ($-\infty$).

2. Find the zeros of the polynomial and plot the x-intercepts.

   p must be a factor of -8 and q must be a factor of -1.

   $$\frac{factors\ of\ 8}{factors\ of\ -1} = \frac{\pm 1, \pm 2, \pm 4, \pm 8}{\pm 1}$$

   Written in lowest terms, $\frac{p}{q} = \pm 1, \pm 2, \pm 4, \pm 8$ is the list of possible rational zeros.

   Now use synthetic division to determine which numbers from the list result in a zero remainder. The results can be summarized in a table.

Zeros	Multiplicity	Cross/Turning Point
$x = -2$	1	Cross
$x = 2$	2	Turning Point

   $\underline{2|}\ -1\ \ \ 2\ \ \ 4\ \ -8$
   $\phantom{\underline{2|}\ -1\ \ \ }-2\ \ \ 0\ \ \ 8$
   $\phantom{\underline{2|}\ }-1\ \ \ 0\ \ \ 4\ \ \ 0$

   $\underline{2|}\ -1\ \ \ 0\ \ \ 4$
   $\phantom{\underline{2|}\ -1\ \ \ }-2\ -4$
   $\phantom{\underline{2|}\ }-1\ -2\ \ \ 0$

   $\underline{-2|}\ -1\ -2$
   $\phantom{\underline{-2|}\ -1\ \ \ }2$
   $\phantom{\underline{-2|}\ }-1\ \ \ 0$

3. Find the turning points, using the relative minimum and relative maximum feature in your calculator.

   The relative maximum is $(2, 0)$ and the relative minimum is $(-0.67, -9.48)$.

4. Find $p(0)$ and plot the y-intercept.

   $p(0) = -8$ $(0, -8)$

5. Plot additional points if necessary to finish the smooth continuous curve.

   Use the Remainder Theorem to find other values.

   $p(-3) = 25$ $p(-1) = -9$ $p(1) = -3$ $p(3) = -5$

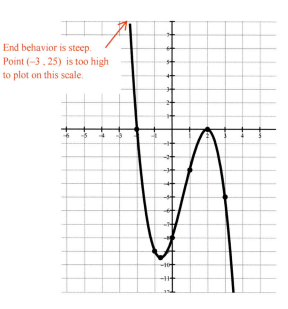

End behavior is steep.
Point $(-3, 25)$ is too high to plot on this scale.

ALGEBRA REVIEW

Solve the quadratic function by completing the square.

1. $x^2 + 4x - 20 = 0$
2. $2x^2 - 3x + 10 = 0$

Find the distance between two points.

$$d = \sqrt{(x_2 - x_1)^2 + (y_2 - y_1)^2}$$

3. $(-3, 5)$ and $(0, -2)$
4. $\left(\dfrac{1}{2}, \dfrac{3}{4}\right)$ and $\left(-\dfrac{5}{2}, \dfrac{1}{4}\right)$

Find the midpoint between two points.

$$\left(\dfrac{x_1 + x_2}{2}, \dfrac{y_1 + y_2}{2}\right)$$

5. $(-3, 5)$ and $(0, -2)$
6. $\left(\dfrac{1}{2}, \dfrac{3}{4}\right)$ and $\left(-\dfrac{5}{2}, \dfrac{1}{4}\right)$

Lesson 4   Practice Exercises

In Exercises 1 – 4, use the leading term test to determine the end-behavior of the graph. Then determine the maximum number of real zeros and turning points of the function.

1. $P(x) = 5x^3 + 7x^2 - x + 9$

2. $P(x) = -2x^5 - x^3 + 3$

3. $P(x) = -11x^4 - 6x^2 + x + 3$

4. $P(x) = x^6 + 3x^5 - 4x^4$

In Exercises 5 – 14, find the zeros of the polynomial. Determine the multiplicity of each zero and the behavior of the graph at each zero.

5. $P(x) = (x-2)(x+1)^2(x-3)^3$

6. $P(x) = (3x+5)^2(x-2)^5(x+8)$

7. $P(x) = (x+7)^3(4x-1)^2(x^2-5)$

8. $P(x) = (4-x)(2x-3)^3(x^2+6)$

9. $P(x) = (x-6)^4(2-x)(x^2+4)$

10. $P(x) = x(x+3)^3(x^2-2)^2$

11. $P(x) = 2x^3 + 5x^2 - x - 6$

12. $P(x) = x^4 + 4x^3 - 2x^2 - 12x + 9$

13. $P(x) = x^4 + 2x^3 - 3x^2 - 8x - 4$

14. $P(x) = x^5 - 6x^4 + 11x^3 - 2x^2 - 12x + 8$

In Exercises 15 – 18, find the turning points using the maximum/minimum function on the calculator. Round your x and y coordinates to the hundredths place, if necessary.

15. $P(x) = 2x^3 - 15x^2 + 24x + 19$

16. $P(x) = 2x^3 + 3x^2 - 12x + 1$

17. $P(x) = x^4 - 6x^3 + 8x^2 + 2x - 1$

18. $P(x) = -2x^4 + 5x^2 + x - 1$

In Exercises 19 – 30, follow the 5 guidelines to graph the polynomial. Be sure to connect the points with a smooth and continuous curve.

19. $P(x) = x^3 + 5x^2 + 2x - 8$

20. $P(x) = -x^3 - 2x^2 + 5x + 6$

21. $P(x) = 2x^3 + x^2 - 8x - 4$

22. $P(x) = x^4 - 7x^3 + 12x^2 + 4x - 16$

23. $P(x) = -x^4 + 9x^2 - 4x - 12$

24. $P(x) = 3x^4 - 11x^3 + 10x - 4$

25. $P(x) = -2x^4 + 3x^3$

26. $P(x) = 2x^5 - 8x^3$

27. $P(x) = -x^4 + 3x^2 + 4$

28. $P(x) = x^4 - 2x^2 + 1$

29. $P(x) = x^7 - 4x^6 + 7x^5 - 12x^4 + 12x^3$

30. $P(x) = 2x^4 - 3x^3 - 15x^2 + 32x - 12$

Solutions to Practice Exercises **Lesson 4**

1. $n = 3$ $a_n = 5$ Far Left: Falls, Far Right: Rises

Maximum number of zeros: 3   Maximum number of turning points: 2

2. $n = 5$ $a_n = -2$ Far Left: Rises, Far Right: Falls

Maximum number of zeros: 5   Maximum number of turning points: 4

3. $n = 4$ $a_n = -11$ Far Left: Falls, Far Right: Falls

Maximum number of zeros: 4   Maximum number of turning points: 3

4. $n = 6$ $a_n = 1$ Far Left: Rises, Far Right: Rises

Maximum number of zeros: 6   Maximum number of turning points: 5

5.
Zeros	Multiplicity	Cross/TP
−1	2	TP
2	1	Cross
3	3	Cross

6.
Zeros	Multiplicity	Cross/TP
−5/3	2	TP
2	5	Cross
−8	1	Cross

7.
Zeros	Multiplicity	Cross/TP
−7	3	Cross
¼	2	TP
$\sqrt{5}$	1	Cross
$-\sqrt{5}$	1	Cross

8.
Zeros	Multiplicity	Cross/TP
4	1	Cross
3/2	3	Cross
$i\sqrt{6}$	1	Neither
$-i\sqrt{6}$	1	Neither

9.
Zeros	Multiplicity	Cross/TP
6	4	TP
2	1	Cross
$2i$	1	Neither
$-2i$	1	Neither

10.
Zeros	Multiplicity	Cross/TP
0	1	Cross
−3	3	Cross
$\sqrt{2}$	2	TP
$-\sqrt{2}$	2	TP

11.
Zeros	Multiplicity	Cross/TP
−2	1	Cross
1	1	Cross
−3/2	1	Cross

12.
Zeros	Multiplicity	Cross/TP
−3	2	TP
1	2	TP

13.
Zeros	Multiplicity	Cross/TP
−2	1	Cross
−1	2	TP
2	1	Cross

14.
Zeros	Multiplicity	Cross/TP
−1	1	Cross
1	1	Cross
2	3	Cross

15. Relative Maximum: $(1,30)$ Relative Minimum: $(4,3)$

16. Relative Maximum: $(-2,21)$ Relative Minimum: $(1,-6)$

17. Relative Maximum: $(1.42, 4.86)$ Relative Minimum: $(-0.11,-1.12)$ and $(3.20,-4.43)$

18. Relative Maximum: $(-1.06,1.03)$ and $(1.17,3.27)$ Relative Minimum: $(-0.10,-1.05)$

19. Far Left: Falls, Far Right: Rises,

    x-int: $(-4,0),(-2,0),(1,0)$, y-int: $(0,-8)$

    Relative Maximum: $(-3.12, 4.06)$

    Relative Minimum: $(-0.21,-8.21)$

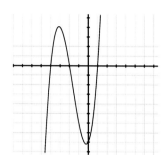

20. Far Left: Rises, Far Right: Falls,

    x-int: $(-3,0),(-1,0),(2,0)$, y-int: $(0,6)$

    Relative Maximum: $(0.79, 8.21)$

    Relative Minimum: $(-2.12,-4.06)$

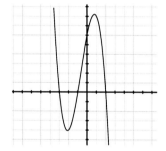

21. Far Left: Falls, Far Right: Rises,

    x-int: $(-2,0),\left(-\dfrac{1}{2},0\right),(2,0)$, y-int: $(0,-4)$

    Relative Maximum: $(-1.33, 3.70)$

    Relative Minimum: $(1,-9)$

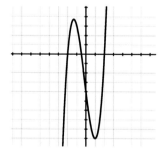

22. Far Left: Rises, Far Right: Rises,

    x-int: $(-1,0),(2,0),(4,0)$, y-int: $(0,-16)$

    Relative Maximum: $(2,0)$

    Relative Minimum: $(-0.15,-16.3)$ and $(3.40,-5.17)$

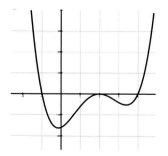

23. Far Left: Falls, Far Right: Falls,

x-int: $(-3,0),(-1,0),(2,0)$, y-int: $(0,-12)$

Relative Maximum: $(-2.22,16.95)$ and $(2,0)$

Relative Minimum: $(0.22,-12.45)$

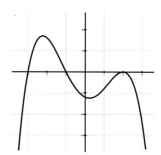

24. Far Left: Rises, Far Right: Rises,

x-int: $(-1,0),\left(\frac{2}{3},0\right),(2\pm\sqrt{2},0)$, y-int: $(0,-4)$

Relative Maximum: $(0.63,0.02)$

Relative Minimum: $(-0.51,-7.44)$ and $(2.63,-34.28)$

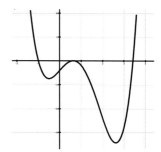

25. Far Left: Falls, Far Right: Falls,

x-int: $(0,0),\left(\frac{3}{2},0\right)$, y-int: $(0,0)$

Relative Maximum: $(1.12,1.07)$

Relative Minimum: none

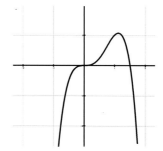

26. Far Left: Falls, Far Right: Rises,

x-int: $(-2,0),(0,0),(2,0)$, y-int: $(0,0)$

Relative Maximum: $(-1.55,11.90)$

Relative Minimum: $(1.55,-11.90)$

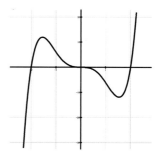

27. Far Left: Falls, Far Right: Falls,

x-int: $(-2,0),(2,0)$, y-int: $(0,4)$

Relative Maximum: $(-1.22,6.25)$ and $(1.22,6.25)$

Relative Minimum: $(0,4)$

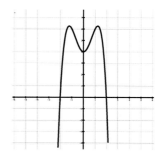

28. Far Left: Rises, Far Right: Rises,

x-int: $(-1,0), (1,0)$, y-int: $(0,1)$

Relative Maximum: $(0,1)$

Relative Minimum: $(-1,0)$ and $(1,0)$

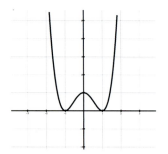

29. Far Left: Falls, Far Right: Rises,

x-int: $(0,0), (2,0)$, y-int: $(0,0)$

Relative Maximum: $(1.30, 5.05)$

Relative Minimum: $(2,0)$

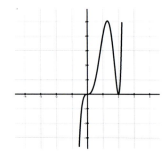

30. Far Left: Rises, Far Right: Rises,

x-int: $(-3,0), \left(\dfrac{1}{2},0\right), (2,0)$, y-int: $(0,-12)$

Relative Maximum: $(1.04, 4.02)$

Relative Minimum: $(2,0)$ and $(-1.92, -80.32)$

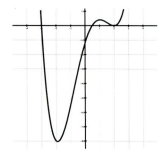

Cumulative Review 1   Lessons 1 – 4

1. Graph the piecewise function and evaluate the function at the given value. Find the domain and range of the function.

   a. $f(x) = \begin{cases} 2x-6 & \text{if } x > 3 \\ 4-x^2 & \text{if } x \leq 3 \end{cases}$

   $f(3)$ and $f(0)$

   b. $g(x) = \begin{cases} 6 & \text{if } x = 2 \\ \dfrac{x^2 - 5x + 6}{x-2} & \text{if } x \neq 2 \end{cases}$

   $g(2)$ and $g(-4)$

   c. $f(x) = \begin{cases} 3 & x \leq -2 \\ |x+1| & -2 < x \leq 0 \\ \sqrt{x}+1 & 0 < x \leq 9 \end{cases}$

   $f(-2)$ and $f(10)$

   d. $h(x) = [[x]] + 1$

   $h(4.78)$ and $h(-6.001)$

2. Write the following absolute functions as piecewise functions.
   a. $f(x) = |6 - 3x|$
   b. $f(x) = |x+3| - |x-7|$

3. Use synthetic division to find the quotient and the remainder:
   a. $(x^3 - 3x + 10) \div (x - 2)$
   b. $(x^4 - 3x^2 - 10) \div (x + 2)$

4. Use the Remainder Theorem to find $P(3)$ for $P(x) = 2x^4 - 9x^2 - 3x + 8$.

5. Determine if $x+1$ or $x-4$ are factors of the polynomial $P(x) = x^3 - 3x^2 - 6x + 8$. Explain your answer using the Factor Theorem.

6. Use the Intermediate value theorem to determine whether $f(x) = 3x^2 - 2x - 11$ has a real zero between $a = 2$ and $b = 3$.

7. Find a polynomial function of lowest degree with integer coefficients and has zeros of $-3, -3i, \sqrt{3}$.

8. Use the Rational Zero Theorem to list all the possible rational zeros for
   a. $f(x) = 6x^4 + 7x^3 - 12x^2 - 3x + 2$
   b. $g(x) = 2x^3 + 7x^2 + 2x - 8$

9. Use Descartes' Rule of Signs for following polynomials and complete the table.
   a. $f(x) = 6x^4 + 7x^3 - 12x^2 - 3x + 2$
   b. $g(x) = 2x^3 + 7x^2 + 2x - 8$

10. Find all the zeros for the following. Show your work using synthetic division.
    a. $f(x) = 6x^4 + 7x^3 - 12x^2 - 3x + 2$
    b. $g(x) = 2x^3 + 7x^2 + 2x - 8$

11. Factor the polynomial into linear factors.
    a. $f(x) = 6x^4 + 7x^3 - 12x^2 - 3x + 2$
    b. $g(x) = 2x^3 + 7x^2 + 2x - 8$

12. Using a graphing calculator, find the relative maxima and minima.
    a. $f(x) = 6x^4 + 7x^3 - 12x^2 - 3x + 2$
    b. $g(x) = 2x^3 + 7x^2 + 2x - 8$

13. Examine the leading term and determine the far-left and far-right behavior of the graph of:
    a. $f(x) = 6x^4 + 7x^3 - 12x^2 - 3x + 2$
    b. $g(x) = 2x^3 + 7x^2 + 2x - 8$

14. Graph the following functions.
    a. $f(x) = 6x^4 + 7x^3 - 12x^2 - 3x + 2$
    b. $g(x) = 2x^3 + 7x^2 + 2x - 8$

15. Find the zeros of $P(x) = (x+4)^3(x-9)^2(x^2-3)$ and state the multiplicity of each zero. Determine the behavior of the graph at each zero.

Solutions to Review

1a.

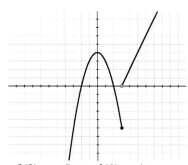

$f(3) = -5 \quad f(0) = 4$
$D:(-\infty,\infty) \quad R:(-\infty,\infty)$

1b.

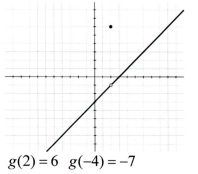

$g(2) = 6 \quad g(-4) = -7$
$D:(-\infty,\infty) \quad R:(-\infty,-1)\cup(-1,\infty)$

1c.

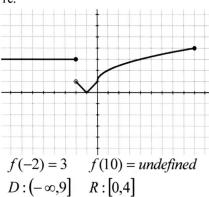

$f(-2) = 3 \quad f(10) = \text{undefined}$
$D:(-\infty,9] \quad R:[0,4]$

1d.

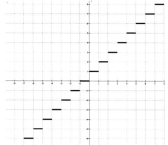

$h(4.78) = 5 \quad h(-6.001) = -6$
$D:(-\infty,\infty) \quad R:\{...-3,-2,-1,0,1,2,3,...\}$

2a. $f(x) = \begin{cases} 6-3x & \text{for } x \leq 2 \\ -6+3x & \text{for } x > 2 \end{cases}$

2b. $f(x) = \begin{cases} -10 & \text{for } x < -3 \\ 2x-4 & \text{for } -3 \leq x < 7 \\ 10 & \text{for } x \geq 7 \end{cases}$

3a. $Q(x) = x^2 + 2x + 1, \quad R(x) = 12$ 3b. $Q(x) = x^3 - 2x^2 + x - 2, R(x) = -6$

4. 80   5. $x+1$ is not a factor of $P(x) = x^3 - 3x^2 - 6x + 8$ since $P(-1) \neq 0$.
$x-4$ is a factor of $P(x) = x^3 - 3x^2 - 6x + 8$ since $P(4) = 0$.

6. yes   7. $P(x) = x^5 + 3x^4 + 6x^3 + 18x^2 - 27x - 81$

8a. $\pm 1, \pm \dfrac{1}{2}, \pm \dfrac{1}{3}, \pm \dfrac{1}{6}, \pm 2, \pm \dfrac{2}{3}$ 8b. $\pm 1, \pm \dfrac{1}{2}, \pm 2, \pm 4, \pm 8$

9a.

P	N	C
2	2	0
2	0	2
0	2	2
0	0	4

9b.

P	N	C
1	2	0
1	0	2

10a. $-2, -\dfrac{1}{2}, \dfrac{1}{3}, 1$ 10b. $-2, \dfrac{-3 \pm \sqrt{41}}{4}$

11a. $f(x) = (x+2)(x-1)(3x-1)(2x+1)$ 11b. $g(x) = (x+2)(2x^2 + 3x - 4)$
$= (x+2)(4x+3-\sqrt{41})(4x+3+\sqrt{41})$

12a. min $(-1.490, -13.754)$ and $(0.730, -2.158)$; max $(-0.115, 2.177)$

12b. min $(-0.153, -8.149)$; max $(-2.180, 0.186)$

13a. up on the left, up on the right   13b. down on the left, up on the right
As $x \to -\infty, y \to \infty$ and as $x \to \infty, y \to \infty$.   As $x \to -\infty, y \to -\infty$ and as $x \to \infty, y \to \infty$.

14a.   14b.   15.

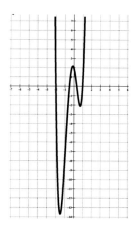

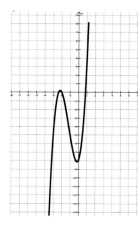

Zeros	Multiplicity	Cross or Turning Point
-4	3	Cross
9	2	Turning Point
$\sqrt{3}$	1	Cross
$-\sqrt{3}$	1	Cross

Lesson 5 — Rational Functions

A **rational function** is defined as $f(x) = \dfrac{p(x)}{q(x)}$, where $p(x)$ and $q(x)$ are polynomials.

The domain of a rational function includes all x values such that $q(x) \neq 0$.

When $q(x) = 0$, the graph of the rational function will be discontinuous. For these x values, a vertical asymptote or a removable discontinuity will exist on the graph. A **vertical asymptote** is the line $x = c$ provided $f(x) \to \infty$ or $f(x) \to -\infty$ as $x \to c$. As the x values approach c, the y values will either increase or decrease without bound. If the numerator and denominator share the common factor $x - c$, a **removable discontinuity** occurs. This factor cancels out, creating a hole in the graph at $x = c$.

Example 1: Determine the domain for the following functions. Then, label the discontinuities as a vertical asymptote or hole.

$$f(x) = \frac{x+3}{x^2 - 9}$$

Solution: The domain of $f(x)$ will include x values for which $x^2 - 9 \neq 0$. Use the square root method to solve for x.

$$x^2 - 9 \neq 0$$
$$x^2 \neq 9$$
$$x \neq \pm 3$$

Written in interval notation, the domain is $(-\infty, -3) \cup (-3, 3) \cup (3, \infty)$. The graph of $f(x)$ is discontinuous at $x = \pm 3$. To determine if the values $x = -3$ and $x = 3$ are vertical asymptotes, look at the factored form of the function.

$$f(x) = \frac{x+3}{x^2 - 9} = \frac{x+3}{(x+3)(x-3)} = \frac{1}{x-3}$$

Notice, the factor $(x-3)$ does not cancel. Therefore, $x = 3$ is a vertical asymptote for the graph of $f(x)$. The function $f(x) \to \pm\infty$ as $x \to 3$. The factor $(x+3)$ cancels, so $x = -3$ will create a hole in the graph for the function. To find the coordinates of the hole, use the reduced form of the function. For $x = -3$, the reduced fraction results in the y value of the hole. $y = \dfrac{1}{(-3)-3} = \dfrac{1}{-6}$

In Lesson 1, we called this hole a removable discontinuity. $x = 3$ is a vertical asymptote and $\left(-3, -\dfrac{1}{6}\right)$ will be a hole in the graph.

A **horizontal asymptote** is the line $y = b$ provided $f(x) \to b$ as $x \to \infty$ or $x \to -\infty$. As x increases or decreases without bound, the y values approach b. From college algebra, you may recall three possibilities for the horizontal asymptote.

Case 1: If the degree of the numerator is less than the degree of denominator, the horizontal asymptote is automatically $y = 0$.

Take a look at Example 1. $\quad f(x) = \dfrac{x+3}{x^2 - 9} = \dfrac{1}{x-3}$ ← Degree 0; no x variable
← Degree 1

The degree of the numerator is less than the denominator, so $y = 0$ is the horizontal asymptote.

Case 2: If the degree of the numerator is equal to the degree of denominator, the horizontal asymptote is the ratio of leading coefficients.

$$y = \frac{\text{leading coefficient of } p(x)}{\text{leading coefficient of } q(x)} \quad \text{See Example 2.}$$

Case 3: If the degree of the numerator is greater than the degree of the denominator, there is no horizontal asymptote. Using long division, an **oblique (or slant) asymptote** can be found.

Example 2: **Find the horizontal or oblique asymptote of the function.**

$$g(x) = \frac{4x^2}{2x^2 + 1}$$

Solution:

The degree of the numerator is 2. The degree of the denominator is 2. Therefore, the horizontal asymptote is $y = \dfrac{\text{leading coefficient of } p(x)}{\text{leading coefficient of } q(x)}$. The horizontal asymptote is $y = 2$.

Much like the Leading Term Test describes the end behavior of the polynomial function, the horizontal asymptote or oblique asymptote describes the end behavior of the rational function. For Example 2, the end behavior of the graph of the function $g(x) = \dfrac{4x^2}{2x^2 + 1}$ will approach 2 as $x \to \infty$ or $x \to -\infty$.

> **Guidelines for Graphing Rational Functions**
>
> 1. Find all asymptotes. Draw these asymptotes on the graph using dotted lines.
>
> 2. Determine if the graph crosses the horizontal or oblique asymptote. Find the coordinates of the crossing point, if applicable.
>
> 3. Find the x and y intercepts.
>
> 4. Plot at least 3 points in between each asymptote.
>
> 5. Draw a smooth curve.

Example 3: Use the 5 guidelines to graph the following functions.

a. $f(x) = \dfrac{x+3}{x^2-9}$

b. $g(x) = \dfrac{4x^2}{2x^2+1}$

Solution:

a. In Example 1, the vertical asymptote and hole was found. The horizontal asymptote will be $y = 0$, since the degree of the numerator is less than the degree of the denominator. We need to determine if the graph will cross the horizontal asymptote. Look at the reduced form. Set it equal to zero and solve for x. If there is a solution, the graph crosses the horizontal asymptote.

$$f(x) = \dfrac{x+3}{x^2-9} = \dfrac{1}{x-3} \qquad\qquad \dfrac{1}{x-3} = 0$$

For the equation, cross multiply. The result, $1 = 0$, is false. Therefore, the graph does not cross the horizontal asymptote.

Find the x-intercepts by solving $f(x) = 0$. Again, use the reduced form. $\dfrac{1}{x-3} = 0$ gives no solution, so there are no x-intercepts for this function.

Find the y-intercepts by evaluating $f(0)$. $\qquad f(0) = \dfrac{1}{0-3} = -\dfrac{1}{3}$

The point $\left(0, -\dfrac{1}{3}\right)$ is the y-intercept.

To plot more points, use the table feature in your calculator.

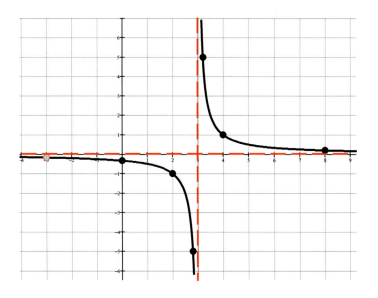

b. Notice the domain of the function is all real numbers $(-\infty, \infty)$. The denominator does not equal to zero for any x value, $2x^2 + 1 \neq 0$. Therefore, $g(x)$ does not have a vertical asymptote or hole in the graph.

In Example 2, the horizontal asymptote was found to be $y = 2$. We need to determine if the graph will cross the horizontal asymptote. Set the function equal to 2 and solve for x. To solve the equation, cross multiply. There is no solution to the equation. Therefore, the graph does not cross the horizontal asymptote.

Cross multiply.
$$\frac{4x^2}{2x^2+1} = 2$$

$$4x^2 = 2(2x^2 + 1)$$
$$4x^2 = 4x^2 + 2$$
$$0 = 2$$

Find the x-intercept by solving $g(x) = 0$.

$$\frac{4x^2}{2x^2+1} = 0$$

$4x^2 = 0$ Multiplicity of 2
$x^2 = 0$
$x = 0$

The x-intercept is the point $(0,0)$. Notice the multiplicity for the factor is 2. Therefore, the graph will have a turning point at the x-intercept. Since $(0,0)$ is also on the y-axis, we have also found the y-intercept.

To plot more points, use the table feature in your calculator. This function is called an even function due to the y-axis symmetry. You can also see that $g(-x) = g(x)$ for all x values. This means less work! For every point you find on the right side, there is a matching point of the left side. Opposite x values result in the same y value. For every point on the graph, you can use the y-axis symmetry to find another point.

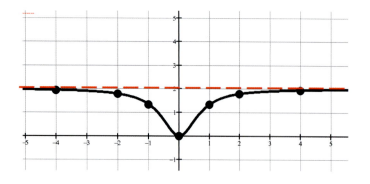
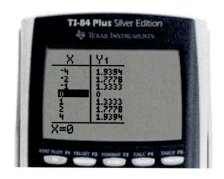

Example 4: Graph the following functions.

a. $h(x) = \dfrac{x^3 - 1}{x^2 - 4}$

b. $r(x) = \dfrac{8x^3 - 1}{2x - 1}$

Solution:

a. The denominator equals zero for $x = \pm 2$. Therefore, the domain of the function is all real numbers except ± 2. $(-\infty, -2) \cup (-2, 2) \cup (2, \infty)$ Look at the factored form of the function to determine whether the function has a vertical asymptote or hole in the graph.

$$h(x) = \dfrac{x^3 - 1}{x^2 - 4} = \dfrac{(x-1)(x^2 + x + 1)}{(x-2)(x+2)}$$

There are no common factors, so nothing will cancel or reduce. This means both x values are vertical asymptotes. The graph never crosses a vertical asymptote due to the domain restrictions.

There is no horizontal asymptote, since the degree of the numerator is greater than the degree of the denominator. This indicates an oblique asymptote is present in this function. To find the equation of the asymptote, perform long division. The quotient is the oblique asymptote. The remainder is discarded.

$$\begin{array}{r} x \\ x^2 - 4 \overline{) x^3 + 0x^2 + 0x - 1} \\ \underline{x^3 - 4x} \\ 4x \end{array}$$

$$y = x$$

To determine if the graph crosses the asymptote, solve the equation $\dfrac{x^3 - 1}{x^2 - 4} = x$. If there is a solution to the equation, then the graph crosses the asymptote.

cross multiply

$\dfrac{x^3 - 1}{x^2 - 4} = x$

$x^3 - 1 = x(x^2 - 4)$

$x^3 - 1 = x^3 - 4x$

$-1 = -4x$

$\dfrac{1}{4} = x$

Therefore, the graph will cross the oblique asymptote at the point $\left(\dfrac{1}{4}, \dfrac{1}{4}\right)$.

Find the x-intercept by solving $h(x) = 0$.

$$\dfrac{x^3 - 1}{x^2 - 4} = 0$$
$$x^3 - 1 = 0$$
$$x^3 = 1$$
$$x = 1$$

Find the y-intercept by evaluating $h(0)$.

$$h(0) = \dfrac{0^3 - 1}{0^2 - 4} = \dfrac{1}{4}$$

The x-intercept is the point $(1,0)$, and $\left(0, \dfrac{1}{4}\right)$ is the y-intercept. To plot more points, use the table feature in your calculator.

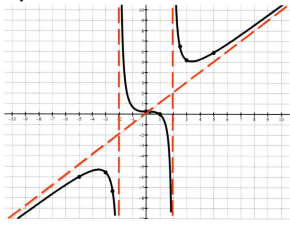
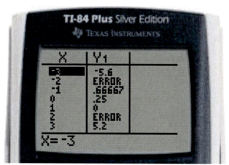

---

b. The denominator equals zero for $x = \dfrac{1}{2}$. Therefore, the domain of the function is all real numbers except $\dfrac{1}{2}, \left(-\infty, \dfrac{1}{2}\right) \cup \left(\dfrac{1}{2}, \infty\right)$. Look at the factored form of the function to determine whether the function has a vertical asymptote or hole in the graph.

Remember SOAP

To factor the difference of two cubes, use the formula. You can also use synthetic division.

$$r(x) = \dfrac{8x^3 - 1}{2x - 1} = \dfrac{(2x - 1)(4x^2 + 2x + 1)}{(2x - 1)} = 4x^2 + 2x + 1$$

$$a^3 - b^3 = (a - b)(a^2 + ab + b^2)$$

The common factor $2x - 1$ cancels resulting in a reduced form of the function. This indicates that $x = \dfrac{1}{2}$ is not a vertical asymptote but rather the location of a hole in the graph. To find the actual coordinates of the hole, evaluate $x = \dfrac{1}{2}$ into the reduced function.

$$y = 4\left(\dfrac{1}{2}\right)^2 + 2\left(\dfrac{1}{2}\right) + 1 = 3$$

The coordinates of the hole is $\left(\dfrac{1}{2}, 3\right)$.

Notice, after reducing the rational function, there is no denominator. Therefore, there is no horizontal or oblique asymptote.

Find the x-intercepts by solving $r(x) = 0$. You can use the reduced form, $4x^2 + 2x + 1 = 0$.

This quadratic equation does not factor, so use the quadratic formula.

$$x = \dfrac{-(2) \pm \sqrt{(2)^2 - 4(4)(1)}}{2(4)} = \dfrac{-2 \pm \sqrt{-12}}{8} = \dfrac{-2 \pm 2i\sqrt{3}}{8} = \dfrac{-1 \pm i\sqrt{3}}{4}$$

Since the solutions are imaginary, there are no real x-intercepts.

Find the y-intercept by evaluating $r(0)$. $\qquad r(0) = 4(0)^2 + 2(0) + 1 = 1$

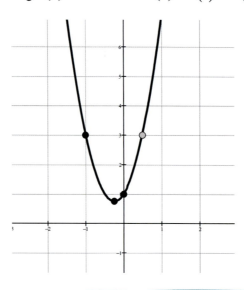

In your calculator, enter the original function and the reduced form. Compare the tables for these two functions.

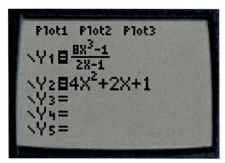

For every x value except for $x = \frac{1}{2}$, the y values agree. You could use a piecewise function from Lesson 1 to repair the hole in the graph to create a continuous function. Recall, $x = \frac{1}{2}$ is called a removable discontinuity.

$$r(x) = \begin{cases} \dfrac{8x^3 - 1}{2x - 1}, & x \neq \dfrac{1}{2} \\ 3, & x = \dfrac{1}{2} \end{cases}$$

In calculus, we need to determine the behavior of the graph at an undefined value. As the value of x approaches $\frac{1}{2}$, what happens to the y-values? One way to determine the behavior of a function is to look at a table of values. Evaluate the function $r(x)$ from Example 4b for values of x close to $\frac{1}{2}$.

x	0.40	0.45	0.49	0.50	0.51	0.55	0.60
y	2.44	2.71	2.94	Error	3.06	3.31	3.64

Another way to determine the behavior is to look at the graph on page 60. With either method, we can see the y-values approach 3. Arrow notation is used in algebra to write the answer.

As $x \to \frac{1}{2}$, $r(x) \to 3$.

In calculus, we will call this a limit. The limit of the function $r(x)$ as x approaches $\frac{1}{2}$ is 3.

$$\lim_{x \to 1/2} r(x) = 3$$

ALGEBRA REVIEW

Solve the inequality and write the solution in interval notation.

1. $-2x + 1 > 5$

2. $\frac{1}{3}x - 4 \leq 8$

3. $-5 < x + 4 < 10$

4. $0 \leq 3 - x \leq 6$

5. $3x \leq -5$ or $3x \geq 5$

6. $2(x+6) < -7$ or $2(x+6) > 7$

Lesson 5   Practice Exercises

In Exercises 1 – 12, find the domain of the rational function. Determine the location of the vertical asymptote and/or hole, if applicable.

1. $f(x) = \dfrac{3x-5}{2x+1}$

2. $f(x) = \dfrac{3}{x-4}$

3. $f(x) = -\dfrac{4x}{x^2+7x+6}$

4. $f(x) = -\dfrac{x^2+x}{2x}$

5. $f(x) = \dfrac{x^2-25}{x-5}$

6. $f(x) = \dfrac{4x^2-11x-3}{4x+1}$

7. $f(x) = \dfrac{x}{x^2+3x}$

8. $f(x) = \dfrac{x+3}{x^2-9}$

9. $f(x) = \dfrac{2x-3}{6x^2-7x-3}$

10. $f(x) = \dfrac{3x+4}{5}$

11. $f(x) = \dfrac{3}{x^2+1}$

12. $f(x) = \dfrac{x-7}{x^2+x+1}$

In Exercises 13 – 24, find the horizontal or oblique asymptote of the rational function. Determine the point where the graph crosses the asymptote, if applicable.

13. $f(x) = \dfrac{3x+5}{x-3}$

14. $f(x) = \dfrac{-4x}{2x+1}$

15. $f(x) = \dfrac{2x+3}{3x^2+7x-6}$

16. $f(x) = \dfrac{-2x}{x^2-1}$

17. $f(x) = \dfrac{-2x^3}{x^2+1}$

18. $f(x) = \dfrac{x^2-x-2}{x-1}$

19. $f(x) = \dfrac{2x^2+7x-4}{x^2+x-2}$

20. $f(x) = \dfrac{3x^2-2x-1}{2x^2+3x-2}$

21. $f(x) = \dfrac{5}{x+1}$

22. $f(x) = \dfrac{x-2}{x^2-x-2}$

23. $f(x) = \dfrac{x^3-x^2-12x}{x^2-7}$

24. $f(x) = \dfrac{3x^3+5x^2+12x-4}{x^2+2x+3}$

In Exercises 25 – 39, graph the rational function using the 5 guidelines.

25. $f(x) = -\dfrac{2}{x+5}$

26. $f(x) = -\dfrac{x}{2x-1}$

27. $f(x) = \dfrac{x-2}{x+3}$

28. $f(x) = \dfrac{x^2-1}{x}$

29. $f(x) = \dfrac{x-2}{x^2+x-6}$

30. $f(x) = \dfrac{x^2+6x+5}{x^2-25}$

31. $f(x) = -\dfrac{8x^3}{x^2-4}$

32. $f(x) = \dfrac{x}{x^2-3x-10}$

33. $f(x) = \dfrac{x-3}{x^2+2x-8}$

34. $f(x) = \dfrac{2x^2}{x^2+9}$

35. $f(x) = -\dfrac{5x}{x^2+1}$

36. $f(x) = \dfrac{x^2+x-12}{x-2}$

37. $f(x) = \dfrac{x^3 - 3x + 2}{x^2 - 9}$

38. $f(x) = \dfrac{2x - 5}{2x^2 + x - 15}$

39. $f(x) = \dfrac{2x^2 - 5x + 5}{x - 2}$

In Exercises 40 – 43, use a piecewise function to repair the hole in the rational function.

40. $f(x) = \dfrac{x^2 - 4}{x^2 - x - 6}$

41. $f(x) = \dfrac{x^2 - 4x - 12}{x - 6}$

42. $f(x) = \dfrac{x^2 - 4x}{x}$

43. $f(x) = \dfrac{x + 4}{x^2 - 2x - 24}$

Solutions to Practice Exercises   Lesson 5

1. $\left(-\infty, -\dfrac{1}{2}\right) \cup \left(-\dfrac{1}{2}, \infty\right)$ VA: $x = -\dfrac{1}{2}$

2. $(-\infty, 4) \cup (4, \infty)$ VA: $x = 4$

3. $(-\infty, -6) \cup (-6, -1) \cup (-1, \infty)$ VA: $x = -6$ and $x = -1$

4. $(-\infty, 0) \cup (0, \infty)$ Hole: $\left(0, -\dfrac{1}{2}\right)$

5. $(-\infty, 5) \cup (5, \infty)$ Hole: $(5, 10)$

6. $\left(-\infty, -\dfrac{1}{4}\right) \cup \left(-\dfrac{1}{4}, \infty\right)$ Hole: $\left(-\dfrac{1}{4}, -\dfrac{13}{4}\right)$

7. $(-\infty, -3) \cup (-3, 0) \cup (0, \infty)$ VA: $x = -3$ Hole: $\left(0, \dfrac{1}{3}\right)$

8. $(-\infty, -3) \cup (-3, 3) \cup (3, \infty)$ VA: $x = 3$ Hole: $\left(-3, -\dfrac{1}{6}\right)$

9. $\left(-\infty, -\dfrac{1}{3}\right) \cup \left(-\dfrac{1}{3}, \dfrac{3}{2}\right) \cup \left(\dfrac{3}{2}, \infty\right)$ VA: $x = -\dfrac{1}{3}$ Hole: $\left(\dfrac{3}{2}, \dfrac{2}{11}\right)$

10. $(-\infty, \infty)$ No VA or Hole

11. $(-\infty, \infty)$ No VA or Hole

12. $(-\infty, \infty)$ No VA or Hole

13. HA: $y = 3$ Doesn't Cross

14. HA: $y = -2$ Doesn't Cross

15. HA: $y = 0$ $\left(-\dfrac{3}{2}, 0\right)$

16. HA: $y = 0$ $(0, 0)$

17. SA: $y = -2x$ $(0, 0)$

18. SA: $y = x$ Doesn't Cross

19. HA: $y = 2$ $(0, 2)$

20. HA: $y = \dfrac{3}{2}$ $\left(\dfrac{4}{13}, \dfrac{3}{2}\right)$

21. HA: $y = 0$ Doesn't Cross

22. HA: $y=0$ $(2,0)$

23. SA: $y=x-1$ $\left(-\dfrac{7}{5},-\dfrac{12}{5}\right)$

24. SA: $y=3x-1$ $\left(\dfrac{1}{5},-\dfrac{2}{5}\right)$

25. VA: $x=-5$, HA: $y=0$

x-int: None, y-int: $\left(0,-\dfrac{2}{5}\right)$

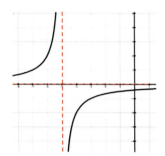

26. VA: $x=\dfrac{1}{2}$, HA: $y=-\dfrac{1}{2}$

x-int: $(0,0)$, y-int: $(0,0)$

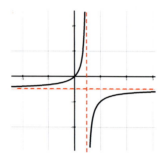

27. VA: $x=-3$, HA: $y=1$

x-int: $(2,0)$, y-int: $\left(0,-\dfrac{2}{3}\right)$

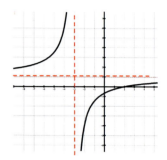

28. VA: $x=0$, SA: $y=x$

x-int: $(\pm 1,0)$, y-int: None

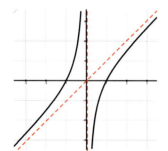

29. VA: $x=-3$, HA: $y=0$

Hole: $\left(2,\dfrac{1}{5}\right)$

x-int: None, y-int: $\left(0,\dfrac{1}{3}\right)$

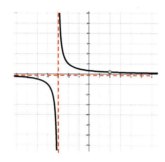

30. VA: $x=5$, HA: $y=1$

Hole: $\left(-5,\dfrac{2}{5}\right)$

x-int: $(-1,0)$, y-int: $\left(0,-\dfrac{1}{5}\right)$

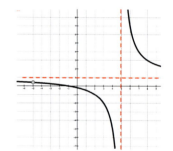

31. VA: $x=\pm 2$, SA: $y=-8x$

x-int: $(0,0)$, y-int: $(0,0)$

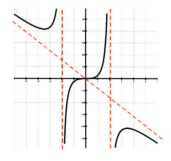

32. VA: $x=-2,5$, HA: $y=0$

x-int: $(0,0)$, y-int: $(0,0)$

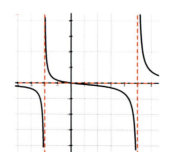

33. VA: $x=-4,2$, HA: $y=0$

x-int: $(3,0)$, y-int: $\left(0,\dfrac{3}{8}\right)$

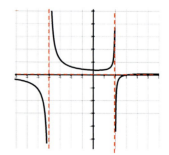

34. VA: None, HA: $y = 2$
x-int: $(0,0)$, y-int: $(0,0)$

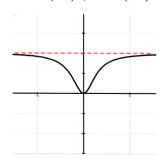

35. VA: None, HA: $y = 0$
x-int: $(0,0)$, y-int: $(0,0)$

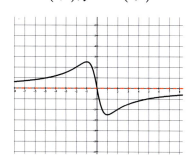

36. VA: $x = 2$, SA: $y = x + 3$
x-int: $(-4,0)\,\&\,(3,0)$, y-int: $(0,6)$

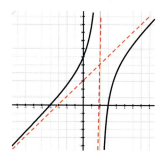

37. VA: $x = \pm 3$, SA: $y = x$

x-int: $(-2,0)\,\&\,(1,0)$, y-int: $\left(0,-\dfrac{2}{9}\right)$

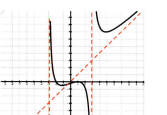

38. VA: $x = -3$, HA: $y = 0$
Hole: $\left(\dfrac{5}{2},\dfrac{2}{11}\right)$

x-int: None, y-int: $\left(0,\dfrac{1}{3}\right)$

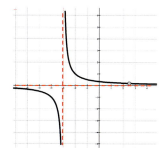

39. VA: $x = 2$, SA: $y = 2x - 1$

x-int: None, y-int: $\left(0,-\dfrac{5}{2}\right)$

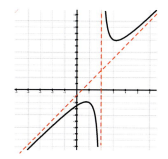

40. $f(x) = \begin{cases} \dfrac{x^2 - 4}{x^2 - x - 6} & \text{for } x \neq -2 \\ \dfrac{4}{5} & \text{for } x = -2 \end{cases}$

41. $f(x) = \begin{cases} \dfrac{x^2 - 4x - 12}{x - 6} & \text{for } x \neq 6 \\ 8 & \text{for } x = 6 \end{cases}$

42. $f(x) = \begin{cases} \dfrac{x^2 - 4x}{x} & \text{for } x \neq 0 \\ -4 & \text{for } x = 0 \end{cases}$

43. $f(x) = \begin{cases} \dfrac{x + 4}{x^2 - 2x - 24} & \text{for } x \neq -4 \\ -\dfrac{1}{10} & \text{for } x = -4 \end{cases}$

Lesson 6 — Inequalities

We studied polynomials functions in Lesson 4 and rational functions in Lesson 5. Now we will use the information to solve polynomial and rational inequalities. When you see the word inequality, a few things should come to mind from previous algebra classes. Do you think of the following symbols?

$\neq$ is not equal to

$<$ is less than

$>$ is greater than

$\leq$ is less than or equal to

$\geq$ is greater than or equal to

Another detail about inequalities to remember: **shading**. When solving inequalities, there isn't just one or two solutions. There is an entire region of solutions. We will write the regions that indicate solutions in interval notation. Don't forget: $<$ and $>$ will use parentheses; $\leq$ and $\geq$ will use brackets.

I. Polynomial Inequalities

Steps to solve polynomial inequalities:

1. Set one side of the inequality equal to zero.

2. Find the zeros of the polynomial. You may use all the techniques used in Lesson 2 & 3. You can also use methods learned in college algebra.

3. Draw a number line to represent the x-axis. Place the zeros on the number line in the correct order.

4. Pick test values in between each zero to determine if the inequality is true or false on the interval.

    You can also use the end behavior and multiplicity of the zeros to determine which intervals are true.

5. The solution set is all the x-values for which the inequality is true. Shade the true intervals and write the solution in interval notation.

Example 1: **Solve the inequality** $2x^4 + x^3 \leq 6x^2 - x - 2$. **Graph the solution and write the solution in interval notation.**

Solution: First, move the terms on the right side of the inequality to the left side to get zero on the right side. $2x^4 + x^3 - 6x^2 + x + 2 \leq 0$

Find the factors of the polynomial by finding the zeros using synthetic division.

```
-2 | 2    1   -6    1    2
   |     -4    6    0   -2
   |_____
     2   -3    0    1    0
```

Continue with synthetic division until you find all the factors.

```
 1 | 2   -3    0    1
   |      2   -1   -1
   |_____
     2   -1   -1    0
```

The quotient is $2x^2 - x - 1$, which is quadratic. We could continue to use synthetic division or factor the quadratic. If it does not factor, the quadratic formula would be needed to find the zeros.

$2x^2 - x - 1$
$(2x+1)(x-1)$

A list of all factors and zeros can be summarized in a table.

Factors	Zeros	Multiplicity	Turning Point or Cross
$x+2$	$x=-2$	1	Cross
$2x+1$	$x=-\frac{1}{2}$	1	Cross
$x-1$	$x=1$	2	Turning Point

Odd multiplicity indicates the sign of $f(x)$ will change from one side of the zero to the other side.

Even multiplicity indicates the sign of $f(x)$ will not change from one side of the zero to the other side.

Draw a number line that represents the x–axis and place the zeros on the number line. These zeros separate the x values into intervals.

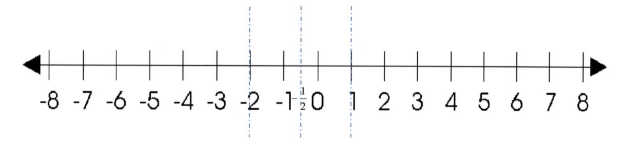

According to the end behavior of the polynomial, the far left rises to positive infinity and the far right also rises to positive infinity. Therefore, the intervals on the far right and far left of the graph are marked with +.

Because the multiplicity of the zeros determine cross or "bounce" behavior, it is concluded that zeros with an odd multiplicity (cross) will create a sign change + to – (or – to + in some cases). For a zero with even multiplicity (turning point), there will be no sign change.

Use this information to determine which intervals are positive and which intervals are negative.

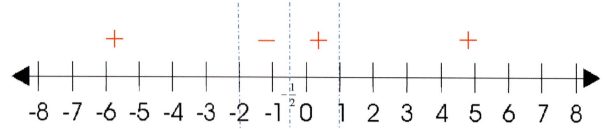

The next step is to determine which of the intervals will make the inequality true. For example 1, $2x^4 + x^3 - 6x^2 + x + 2 \leq 0$. Therefore, the only interval that is part of the solution is between -2 and $-\frac{1}{2}$. What about the zeros? Since the inequality includes less than **or equal to**, brackets are used to indicate the zeros are included in the solution. The zero $x = 1$ would also need to be included. The final solution in interval notation is $\left[-2, -\frac{1}{2}\right] \cup \{1\}$. Now, show the solution on the number line by shading the interval between -2 and $-\frac{1}{2}$, putting brackets at the zeros. Also, put a closed dot at $x = 1$.

Example 2: Solve the inequality $x^5 + 24 > 3x^3 + 8x^2$. **Graph the solution and write the solution in interval notation.**

Solution: First, move the terms on the right side of the inequality to the left side to get zero on the right side. $\quad x^5 - 3x^3 - 8x^2 + 24 > 0$

Now, find the zeros of the polynomial. You may use synthetic division as in example 1, but don't forget the factoring methods learned in algebra. This polynomial will factor using grouping.

$x^5 - 3x^3 - 8x^2 + 24$ Group the first two terms together and notice the GCF is x^3. Group the second two terms together and notice the GCF is -8. Factor by grouping only works if the terms in the parentheses are equal.

$x^3(x^2 - 3) - 8(x^2 - 3)$

$(x^2-3)(x^3-8)$
$(x^2-3)(x-2)(x^2+2x+4)$

There is a formula that may need to be reviewed: Difference of Two Cubes.

Set each factor to zero and solve for x.

$x^2-3=0$
$x^2=3$
$x=\pm\sqrt{3}$

$x-2=0$
$x=2$

$x^2+2x+4=0$
$x=\dfrac{-2\pm\sqrt{(2)^2-4(1)(4)}}{2(1)}$
$x=\dfrac{-2\pm\sqrt{-12}}{2}=-1\pm i\sqrt{3}$

We have 3 real zeros and 2 imaginary zeros. We need to draw a number line and place the real zeros in order. Notice, the number line is separated into 4 intervals. We can use the end behavior and multiplicities of the zeros to determine whether the intervals are + or −. Another method is to pick test values in each interval. Select any x value between the zeros to evaluate in the polynomial. We will have to be creative with interval between $\sqrt{3}$ and 2.

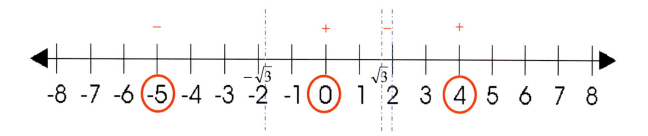

Test Values	Evaluate for the polynomial inequality.		
$x=-5$	$(-5)^5-3(-5)^3-8(-5)^2+24>0$	$-2926>0$	FALSE
$x=0$	$(0)^5-3(0)^3-8(0)^2+24>0$	$24>0$	TRUE
$x=1.9$	$(1.9)^5-3(1.9)^3-8(1.9)^2+24>0$	$-0.69601>0$	FALSE
$x=4$	$(4)^5-3(4)^3-8(4)^2+24>0$	$728>0$	TRUE

Shade the intervals that result in a true inequality. Write the solution in interval notation. Make sure to use parentheses for the zeros.

Solution in interval notation: $\left(-\sqrt{3},\sqrt{3}\right)\cup(2,\infty)$

II. Rational Inequalities

Steps to solve rational inequalities:

1. Set one side of the inequality equal to zero.

2. Find the zeros of the numerator and denominator.

3. Draw a number line to represent the x-axis. Place the zeros on the number line in the correct order.

4. Pick test values in between each zero to determine if the inequality is true or false on the interval.

5. The solution set is all the x-values for which the inequality is true. Shade the true intervals and write the solution in interval notation.

**Note: The denominator zeros will NEVER be included in the solution due to the domain restrictions.

Example 3: Solve the inequality $\dfrac{x^2 + 4x - 12}{x^2 - 4} \geq 0$. Graph the solution and write the solution in interval notation.

Solution: Since the inequality has zero on the right side, proceed to step 2. Find the zeros of the numerator and denominator.

$x^2 + 4x - 12 = 0$ $\qquad$ $x^2 - 4 = 0$
$(x+6)(x-2) = 0$ $\qquad$ $(x+2)(x-2) = 0$
$x = -6 \ or \ x = 2$ $\qquad$ $x = -2 \ or \ x = 2$

Place all zeros on the number line, which separates the x values into intervals. Notice, the factor $(x-2)$ is in both the numerator and denominator. Recall from Lesson 5, there will be a hole in the graph at $x = 2$. This also indicates $(x-2)$ has a multiplicity of 2, and there will be no sign change at $x = 2$. The other two factors have multiplicity of 1, so there will be a sign change at $x = -6$ and $x = -2$. Select a test value between each zero to determine in the inequality is true.

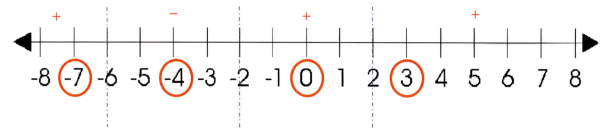

Test Values   Evaluate for the polynomial inequality.

$x = -7$ $\dfrac{(-7)^2 + 4(-7) - 12}{(-7)^2 - 4} \geq 0$ $\dfrac{1}{5} \geq 0$ TRUE

$x = -4$ $\dfrac{(-4)^2 + 4(-4) - 12}{(-4)^2 - 4} \geq 0$ $-1 \geq 0$ FALSE

$x = 0$ $\dfrac{(0)^2 + 4(0) - 12}{(0)^2 - 4} \geq 0$ $3 \geq 0$ TRUE

$x = 3$ $\dfrac{(3)^2 + 4(3) - 12}{(3)^2 - 4} \geq 0$ $\dfrac{9}{5} \geq 0$ TRUE

Shade the intervals that result in a true inequality, and write the solution in interval notation. Notice the zero for the numerator (-6) receives a bracket, and the zeros of the denominator (-2 and 2) receive parentheses.

$(-\infty, -6] \cup (-2, 2) \cup (2, \infty)$

Example 4: Solve the inequality $\dfrac{2x}{x+1} < 1$.

Graph the solution and write the solution in interval notation.

Solution: The first step is to subtract 1 to get zero on the right side of the inequality.

$\dfrac{2x}{x+1} - 1 < 0$

$\dfrac{2x}{x+1} - \dfrac{1(x+1)}{x+1} < 0$

To find the zeros of the numerator and denominator, combine the fractions by getting a common denominator. The LCD is $x+1$.

$$\frac{2x-(x+1)}{x+1}<0$$

$$\frac{x-1}{x+1}<0$$

There is one zero for the numerator, $x=1$, and one zero for the denominator, $x=-1$. Draw a number line and determine which intervals make the inequality true.

Because the degree in the numerator is equal to the degree in the denominator, the horizontal asymptote is $y=1$. This indicates the end behavior. On the far left, the graph approaches 1. This is positive and therefore makes the inequality false. On the far right, the graph approaches 1 also, which makes the inequality false. Since both zeros have multiplicity of 1, there is a sign change at $x=-1$ and $x=1$.

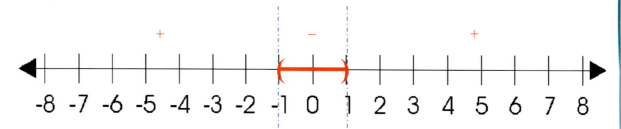

The solution in interval notation is $(-1,1)$.

Example 5: Solve the inequality $\frac{4}{x-2} \le \frac{3}{x-1}$.

Graph the solution and write the solution in interval notation.

Solution: The first step is to subtract $\frac{3}{x-1}$ to get zero on the right side of the inequality.

$$\frac{4}{x-2}-\frac{3}{x-1}\le 0$$

$$\frac{4(x-1)}{(x-2)(x-1)}-\frac{3(x-2)}{(x-1)(x-2)}\le 0$$

$$\frac{4x-4}{(x-2)(x-1)}-\frac{3x-6}{(x-2)(x-1)}\le 0$$

$$\frac{x+2}{(x-2)(x-1)}\le 0$$

> Warning: Don't cross multiply. For inequalities, the sign is switched if both sides are multiplied by a negative. Since x represents any number, how do you know if you should switch the inequality sign? You don't! So, don't multiply both sides of an inequality by a variable expression like $x-2$ and $x-1$.

The zero of the numerator is $x=-2$, and the zeros of the denominator are $x=2$ and $x=1$. Start with one test value. The easiest test value for this problem is at $x=0$.

Evaluate the inequality for $x = 0$, and the result is positive. $\dfrac{0+2}{(0-2)(0-1)} = 1$

Now use the fact there will be a sign change at each zero since they all have multiplicity of 1 (odd multiplicity means sign change). The solution to the inequality includes the intervals that are less than zero. We also have to include the zeros of the numerator. However, the zeros of the denominator are never included due to the domain restrictions.

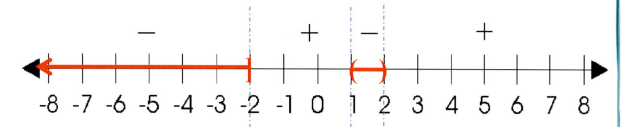

The solution in interval notation is $(-\infty, -2] \cup (1, 2)$.

ALGEBRA REVIEW

Simplify the rational expressions.

1. $\dfrac{5x}{x+2} - 3$

2. $\dfrac{2}{x-4} + 1$

3. $\dfrac{2}{x+1} + \dfrac{1}{x-1}$

4. $\dfrac{1}{x} - \dfrac{2}{(x+1)^2}$

5. $\dfrac{3}{x^2 + 2x - 8} - \dfrac{2}{x^2 - 3x + 2}$

6. $\dfrac{x-1}{x^2 - 2x + 1} - \dfrac{x+1}{x-1}$

Lesson 6

Practice Exercises

In Exercises 1 – 21, solve each polynomial inequality. Graph the solution on a number line and write the solution in interval notation.

1. $x^2 + 6x - 7 \geq 0$
2. $x^2 - 9x - 36 < 0$
3. $x^2 + x > 0$
4. $5x^3 + 10x^2 \leq 0$
5. $2x^2 - 6x \leq -1$
6. $3x^2 - 3x > 4$
7. $4x^2 \geq 2x + 7$
8. $x^2 < -5x - 3$
9. $(x-1)^2(3x+1) \leq 0$
10. $(x+1)(2x-3)^2 \leq 0$
11. $(x-4)(x+2)(x-3)^2 \geq 0$
12. $(5x-1)^2(x-2)(x+1) \geq 0$
13. $2x^3 - 3x^2 - 11x + 6 < 0$
14. $15x^3 + 14x^2 - 3x - 2 > 0$
15. $x^3 + x^2 > 4x + 4$
16. $x^3 + 3x^2 \leq x + 3$
17. $x^4 - 3x^2 + 8 \leq 4x^3 - 10x$
18. $x^4 - 6x^3 > -8x^2 - 6x + 9$
19. $x^3 + x^2 > 5x - 3$
20. $-x^4 + 5x^3 - 4x^2 + 3 < x^3 + 3$
21. $x^4 + 4x - 3 \geq 9x^2 - 15$

In Exercises 22 – 44, solve each rational inequality. Graph the solution on a number line and write the solution in interval notation.

22. $\dfrac{x-2}{x+5} < 0$
23. $\dfrac{x}{x-6} > 0$
24. $\dfrac{x^2}{x-1} \geq 0$
25. $\dfrac{x^2}{2x+5} \leq 0$

26. $\dfrac{x^2-4}{x} > 0$
27. $\dfrac{x+5}{x^2-x} < 0$
28. $\dfrac{x^2-x-6}{2x^2-x-3} \leq 0$
29. $\dfrac{3x^2+x-4}{x^2-3x-10} \leq 0$

30. $\dfrac{x+3}{x-2} \leq 2$
31. $\dfrac{x+4}{2x-1} \geq 3$
32. $\dfrac{x+1}{2x-1} > 1$
33. $\dfrac{1}{x-3} < 1$

34. $\dfrac{x}{x+2} \geq 2$
35. $\dfrac{x+1}{x-2} > 3$
36. $\dfrac{3x+1}{x+2} < \dfrac{3}{2}$
37. $\dfrac{x-2}{x+3} \leq 4$

38. $\dfrac{1}{x+2} > \dfrac{1}{x-3}$
39. $\dfrac{1}{x+1} \leq \dfrac{2}{x-1}$
40. $\dfrac{x-4}{x+3} \leq \dfrac{x+2}{x-1}$
41. $\dfrac{2x-1}{x+3} \geq \dfrac{x+1}{3x+1}$

42. $\dfrac{x+5}{x-4} < \dfrac{3x+2}{2x+1}$
43. $\dfrac{x}{x^2+6x+8} > \dfrac{4}{x^2-4}$
44. $\dfrac{2}{x^2-4x+3} > \dfrac{5}{x^2-9}$

Solutions to Practice Exercises **Lesson 6**

1. $(-\infty, -7] \cup [1, \infty)$

2. $(-3, 12)$

3. $(-\infty, -1) \cup (0, \infty)$

4. $(-\infty, -2] \cup \{0\}$

5. $\left[\dfrac{3-\sqrt{7}}{2}, \dfrac{3+\sqrt{7}}{2}\right]$

6. $\left(-\infty, \dfrac{3-\sqrt{57}}{6}\right) \cup \left(\dfrac{3+\sqrt{57}}{6}, \infty\right)$

7. $\left(-\infty, \dfrac{1-\sqrt{29}}{4}\right] \cup \left[\dfrac{1+\sqrt{29}}{4}, \infty\right)$

8. $\left(\dfrac{-5-\sqrt{13}}{2}, \dfrac{-5+\sqrt{13}}{2}\right)$

9. $\left(-\infty, -\dfrac{1}{3}\right] \cup \{1\}$

10. $(-\infty, -1] \cup \left\{\dfrac{3}{2}\right\}$

11. $(-\infty, -2] \cup \{3\} \cup [4, \infty)$

12. $(-\infty, -1] \cup \left\{\dfrac{1}{5}\right\} \cup [2, \infty)$

13. $(-\infty, -2) \cup \left(\dfrac{1}{2}, 3\right)$

14. $\left(-1, -\dfrac{1}{3}\right) \cup \left(\dfrac{2}{5}, \infty\right)$

15. $(-2, -1) \cup (2, \infty)$

16. $(-\infty, -3] \cup [-1, 1]$

17. $\{-1\} \cup [2,4]$

18. $(-\infty,-1) \cup (1,3) \cup (3,\infty)$

19. $(-3,1) \cup (1,\infty)$

20. $(-\infty,0) \cup (0,2) \cup (2,\infty)$

21. $(-\infty,-3] \cup [-1,\infty)$

22. $(-5,2)$

23. $(-\infty,0) \cup (6,\infty)$

24. $\{0\} \cup (1,\infty)$

25. $\left(-\infty,-\dfrac{5}{2}\right) \cup \{0\}$

26. $(-2,0) \cup (2,\infty)$

27. $(-\infty,-5) \cup (0,1)$

28. $[-2,-1) \cup \left(\dfrac{3}{2},3\right]$

29. $\left(-2,-\dfrac{4}{3}\right] \cup [1,5)$

30. $(-\infty,2) \cup [7,\infty)$

31. $\left(\dfrac{1}{2},\dfrac{7}{5}\right]$

32. $\left(\dfrac{1}{2},2\right)$

33. $(-\infty,3) \cup (4,\infty)$

34. $[-4,-2)$

35. $\left(2,\dfrac{7}{2}\right)$

36. $\left(-2,\dfrac{4}{3}\right)$

37. $\left(-\infty,-\dfrac{14}{3}\right]\cup(-3,\infty)$

38. $(-2,3)$

39. $[-3,-1)\cup(1,\infty)$

40. $\left(-3,-\dfrac{1}{5}\right]\cup(1,\infty)$

41. $(-\infty,-3)\cup\left[\dfrac{5-\sqrt{105}}{10},-\dfrac{1}{3}\right)\cup\left[\dfrac{5+\sqrt{105}}{10},\infty\right)$

42. $\left(-\infty,\dfrac{21-\sqrt{493}}{2}\right)\cup\left(-\dfrac{1}{2},4\right)\cup\left(\dfrac{21+\sqrt{493}}{2},\infty\right)$

43. $(-4,-2)\cup(-2,2)\cup(8,\infty)$

44. $(-3,1)\cup\left(3,\dfrac{11}{3}\right)$

FINAL EXAM REVIEW

KEEP IT FRESH

Use the piece wise function $f(x)=\begin{cases}\dfrac{1}{2}x+1, & -6<x<0 \\ (x-1)^2, & 0\le x<3 \\ -3, & x\ge 3\end{cases}$ to find the following.

1. Evaluate $f(-4)$ and $f(3)$.

2. Graph the function.

3. What is the domain?

4. What is the range?

5. Is the function continuous at $x=0$? At $x=3$?

Lesson 7   Exponential and Logarithmic Functions

Many real world applications use a model that shows exponential growth or decay. Population is one example, and investing money in an account paying interest is another example.

> **Exponential Function** $f(x) = b^x$ where the base of the exponent b is never negative and cannot equal one $b > 0, b \neq 1$
>
> The exponent x must be a real number.

Example 1: Graph the exponential function $f(x) = 2^x$.

Solution: Make a table of values and plot the points. Notice, the domain of the function is $(-\infty, \infty)$ and the range is $(0, \infty)$. There is a horizontal asymptote at $y = 0$.

x	y
3	8
2	4
1	2
0	1
−1	$\frac{1}{2}$
−2	$\frac{1}{4}$
−3	$\frac{1}{8}$

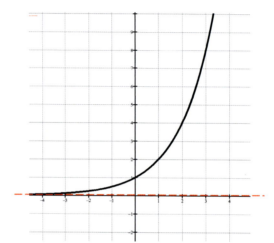

The exponential function is a **one–to–one function**, meaning a one to one correspondence for x and y values. In other words, the x values don't repeat because the graph passes the vertical line test AND the y values don't repeat because the graph passes the horizontal line test. Because $f(x) = b^x$ is one–to–one, we can find the inverse. Follow the steps found in algebra.

Replace $f(x)$ with y. $y = b^x$

Switch the variables x and y. $x = b^y$

Solve for y. This requires a function that will find the exponent, which is defined as a logarithmic function.

> **Logarithmic Function** $y = \log_b x$ is equivalent to $x = b^y$.
>
> y represents the exponent on b that yields x. The value of x must be greater than 0.

Repeat this statement over and over, until it sticks - A logarithm is an exponent! When you evaluate a logarithmic expression, the answer is an exponent.

In college algebra, you learned a few properties of inverses.

1. One–to–One

2. If (a,b) is a point on $f(x)$, then (b,a) is a point on $f^{-1}(x)$.

3. Graphs of $f(x)$ and $f^{-1}(x)$ are symmetric about the line $y = x$.

4. Domain and range switch.

5. $f(f^{-1}(x)) = x$ and $f^{-1}(f(x)) = x$

In the next two examples, notice the inverse properties for the exponential and logarithmic functions.

Example 2: **Graph the exponential function $g(x) = \log_2 x$.**

Solution: Switch the x and y values of the table for $f(x) = 2^x$. The graphs are symmetric about the line $y = x$.

x	y
8	3
4	2
2	1
1	0
$\frac{1}{2}$	−1
$\frac{1}{4}$	−2
$\frac{1}{8}$	−3

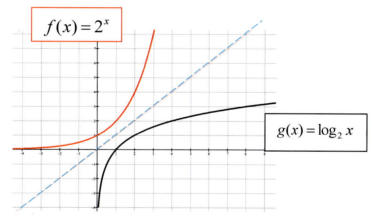

Notice, the domain and range for $f(x) = 2^x$ and $g(x) = \log_2 x$ switch.

	Domain	Range
$f(x) = 2^x$	$(-\infty, \infty)$	$(0, \infty)$
$g(x) = \log_2 x$	$(0, \infty)$	$(-\infty, \infty)$

Also, the logarithmic function $g(x) = \log_2 x$ has a vertical asymptote at $x = 0$.

Example 3: Find the inverse of each function given. Find the domain and range of the function and the inverse.

a. $f(x) = -2^x + 1$ **b.** $g(x) = \dfrac{1}{2} \log_2(x+5)$

Solutions:

a.

		$f(x) = -2^x + 1$
Replace $f(x)$ with y.		$y = -2^x + 1$
Switch the variables x and y.		$x = -2^y + 1$
Solve for y.		$x - 1 = -2^y$
		$-(x-1) = 2^y$
		$\log_2(-x+1) = y$
Replace y with the function $f^{-1}(x)$.		$f^{-1}(x) = \log_2(1-x)$

	Domain	Range
$f(x) = -2^x + 1$	$(-\infty, \infty)$	$(-\infty, 1)$
$f^{-1}(x) = \log_2(-x+1)$	$(-\infty, 1)$	$(-\infty, \infty)$

---

b.

	$g(x) = \dfrac{1}{2} \log_2(x+5)$
Replace $g(x)$ with y.	$y = \dfrac{1}{2} \log_2(x+5)$
Switch the variables x and y.	$x = \dfrac{1}{2} \log_2(y+5)$
Solve for y.	$2x = \log_2(y+5)$
	$2^{2x} = y + 5$
	$2^{2x} - 5 = y$
Replace y with function $g^{-1}(x)$.	$g^{-1}(x) = 2^{2x} - 5$

	Domain	Range
$g(x) = \dfrac{1}{2}\log_2(x+5)$	$(-5, \infty)$	$(-\infty, \infty)$
$g^{-1}(x) = 2^{2x} - 5$	$(-\infty, \infty)$	$(-5, \infty)$

We can have the same transformations studied in Lesson 1 for exponential and logarithmic functions.

Horizontal Shifts	$f(x+c)$	b^{x+c}	$\log_b(x+c)$
Reflections y-axis x-axis	$f(-x)$ $-f(x)$	b^{-x} $-b^x$	$\log_b(-x)$ $-\log_b x$
Vertical Shrink/Stretch	$cf(x)$	$c(b^x)$	$c\log_b x$
Horizontal Shrink/Stretch	$f(cx)$	b^{cx}	$\log_b(cx)$
Vertical Shifts	$f(x)+c$	b^x+c	$\log_b x + c$

Example 4: Graph the function. Be sure to include the asymptote. State the basic function and what transformations are applied.

a. $f(x) = -2^x + 1$ b. $g(x) = \dfrac{1}{2}\log_2(x+5)$

Solutions:

a. The basic function is $y = 2^x$. The graph of $y = 2^x$ is reflected across the x-axis and shifted up one unit to obtain the graph of $f(x) = -2^x + 1$.

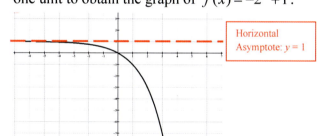

Horizontal Asymptote: $y = 1$

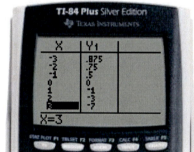

b. The basic function is $y = \log_2 x$. The graph of $y = \log_2 x$ is shifted left five units and has a vertical shrink to obtain the graph of $g(x) = \dfrac{1}{2}\log_2(x+5)$.

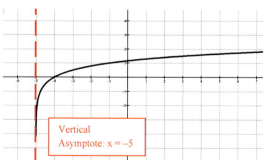

Vertical Asymptote: $x = -5$

When graphing logarithms, some students forget there are values between –5 and –4 that should be plotted. The domain is $(-5, \infty)$. The table on the calculator can be misleading, because it does not show the decimal values. Find the value of the function for $x = -4.5$ and $x = -4.9$. These values indicate the graph continues pass the point $(-4, 0)$.

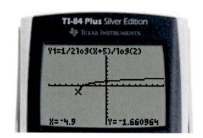

There are two important bases we need to know. Your calculator is programed for these two bases.

Common Base: 10

Graph the following functions.
$f(x) = 10^x$ and $g(x) = \log x$

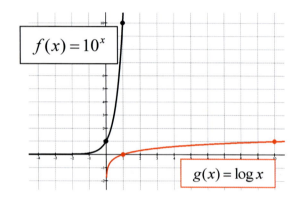

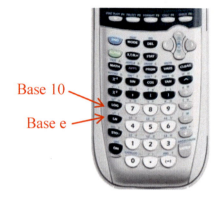

Base 10
Base e

Notice, these functions are symmetric about the line $y = x$. The domain and range switch.

	Domain	Range
$f(x) = 10^x$	$(-\infty, \infty)$	$(0, \infty)$
$g(x) = \log x$	$(0, \infty)$	$(-\infty, \infty)$

Example 5: The level of sound β (in decibels) with an intensity of I is represented by the equation, $\beta = 10 \log \left(\dfrac{I}{I_0} \right)$. I_0 is an intensity of 10^{-16} watt per square centimeter, corresponding roughly to the faintest sound that can be heard.

a. Find the level of sound equal to an intensity of a whisper ($I = 10^{-14}$).
b. Find the level of sound equal to the threshold of pain ($I = 10^{-4}$).

Solution: This is a common logarithm with base 10.

a. $\beta = 10 \log \left(\dfrac{10^{-14}}{10^{-16}} \right)$

$= 10 \log (10^2)$
$= 10(2)$
$= 20$
20 decibels

b. $\beta = 10 \log \left(\dfrac{10^{-4}}{10^{-16}} \right)$

$= 10 \log (10^{12})$
$= 10(12)$
$= 120$
120 decibels

Natural Base: e

The value of $e \approx 2.7182818284$. Where did this natural base come from? Evaluate the expression $\left(1+\dfrac{1}{n}\right)^n$ for $n = 1, 10, 100, 1000, etc$. In other words, as the value of n increases, what happens to the expression?

As $n \to \infty$, the value of the expression $\left(1+\dfrac{1}{n}\right)^n \to e$.

n	$\left(1+\dfrac{1}{n}\right)^n$
1	2
10	2.5937
100	2.7048
1000	2.7169
10000	2.7181
100000	2.7183

Graph the following functions.
$f(x) = e^x$ and $g(x) = \ln x$

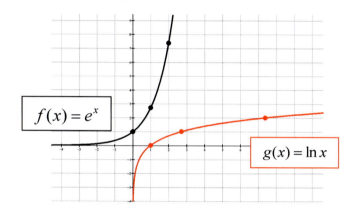

Notice $2 < e < 3$, so the graph of $f(x) = e^x$ is between the graphs of 2^x and 3^x.

$\ln x$ is a logarithm with base e. In other words, $\log_e x = \ln x$. This logarithm is called the natural logarithm.

Example 6: An experimental population of fruit flies increases according to the exponential equation $y = 33e^{0.5493t}$. How many fruit flies will be present after 2 days? After 4 days?

Solution:

$y = 33e^{0.5493(2)}$

$= 98.99878$

≈ 99 fruit flies

$y = 33e^{0.5493(4)}$

$= 296.9927$

≈ 297 fruit flies

83

Lesson 7   Practice Exercises

In Exercises 1 – 6, find the inverse of each function given. Find the domain and range of the function and the inverse.

1. $f(x) = 2^{x-1} + 3$
2. $f(x) = 2e^x - 1$
3. $f(x) = 10^{2x}$

4. $f(x) = 1 - \log_3(x+2)$
5. $f(x) = \dfrac{1}{2}\ln(x-3)$
6. $f(x) = \log(2x-1) + 3$

In Exercises 7 – 24, graph the function. Be sure to include the asymptote. State the basic function and what transformations are applied.

7. $f(x) = 2^{x+1}$
8. $f(x) = 3^{x-2} + 1$
9. $f(x) = \left(\dfrac{1}{2}\right)^x - 3$

10. $f(x) = -\left(\dfrac{1}{4}\right)^{x-1}$
11. $f(x) = e^{x-5} + 4$
12. $f(x) = e^{-x} - 2$

13. $f(x) = \dfrac{1}{2}e^{x+1} + 1$
14. $f(x) = \log(x+1) - 5$
15. $f(x) = 1 - \log(x-4)$

16. $f(x) = 3 + \log_2(x-2)$
17. $f(x) = \log_2(x-1) - 4$
18. $f(x) = 2 - \log_3(x+5)$

19. $f(x) = 2\log_3 x - 6$
20. $f(x) = 3\ln(x+2)$
21. $f(x) = \ln(6-x)$

22. $f(x) = \ln x + 1$
23. $f(x) = \dfrac{1}{2}\ln(x+5) + 1$
24. $f(x) = 2 - \ln(x+1)$

Solutions to Practice Exercises   Lesson 7

1. $f^{-1}(x) = 1 + \log_2(x-3)$
2. $f^{-1}(x) = \ln\left(\dfrac{x+1}{2}\right)$
3. $f^{-1}(x) = \dfrac{1}{2}\log x$

	$f(x)$	$f^{-1}(x)$
Domain	$(-\infty, \infty)$	$(3, \infty)$
Range	$(3, \infty)$	$(-\infty, \infty)$

	$f(x)$	$f^{-1}(x)$
Domain	$(-\infty, \infty)$	$(-1, \infty)$
Range	$(-1, \infty)$	$(-\infty, \infty)$

	$f(x)$	$f^{-1}(x)$
Domain	$(-\infty, \infty)$	$(0, \infty)$
Range	$(0, \infty)$	$(-\infty, \infty)$

4. $f^{-1}(x) = 3^{1-x} - 2$

5. $f^{-1}(x) = e^{2x} + 3$

6. $f^{-1}(x) = \dfrac{1}{2}(10^{x-3} + 1)$

	$f(x)$	$f^{-1}(x)$
Domain	$(-2, \infty)$	$(-\infty, \infty)$
Range	$(-\infty, \infty)$	$(-2, \infty)$

	$f(x)$	$f^{-1}(x)$
Domain	$(3, \infty)$	$(-\infty, \infty)$
Range	$(-\infty, \infty)$	$(3, \infty)$

	$f(x)$	$f^{-1}(x)$
Domain	$\left(\dfrac{1}{2}, \infty\right)$	$(-\infty, \infty)$
Range	$(-\infty, \infty)$	$\left(\dfrac{1}{2}, \infty\right)$

7. Transforms the basic function 2^x by shifting left one unit.

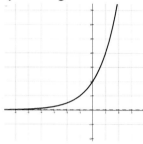

8. Transforms the basic function 3^x by shifting right two units and up one unit.

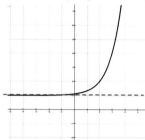

9. Transforms the basic function $\left(\dfrac{1}{2}\right)^x$ by shifting down three units.

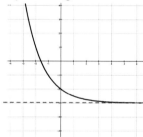

10. Transforms the basic function $\left(\dfrac{1}{4}\right)^x$ by shifting right one unit and reflecting the x-axis.

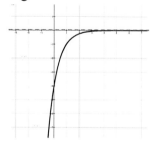

11. Transforms the basic function e^x by shifting right five and up four units.

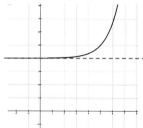

12. Transforms the basic function e^x by shifting down two units and reflecting the y-axis.

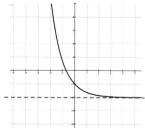

13. Transforms the basic function e^x by shifting left one, up one, and vertical shrink by $\dfrac{1}{2}$.

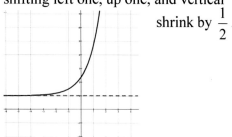

14. Transforms the basic function $\log x$ by shifting left one unit and down five units.

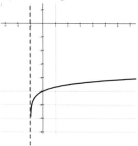

15. Transforms the basic function $\log x$ by shifting right four, up one, and reflect x-axis.

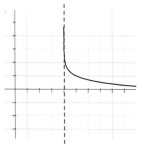

16. Transforms the basic function $\log_2 x$ by shifting right two units and up three units.

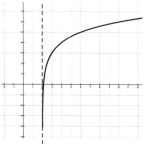

17. Transforms the basic function $\log_2 x$ by shifting right one unit and down four units.

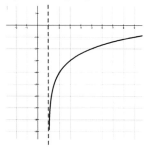

18. Transforms the basic function $\log_3 x$ by shifting left five, up two, and reflect x-axis

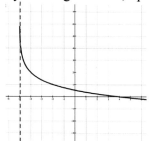

19. Transforms the basic function $\log_3 x$ by shifting down six units and vertical stretch by 2.

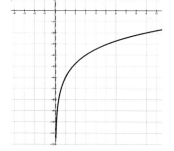

20. Transforms the basic function $\ln x$ by shifting left two units and vertical stretch by 3.

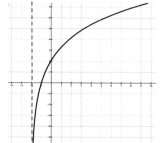

21. Transforms the basic function $\ln x$ by shifting right six units and reflect y-axis.

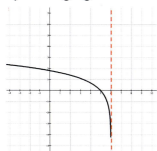

22. Transforms the basic function $\ln x$ by shifting up one unit.

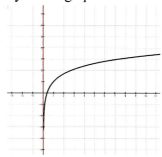

23. Transforms the basic function $\ln x$ by shifting left five, up one, and vertical shrink by $\frac{1}{2}$.

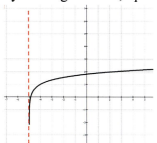

24. Transforms the basic function $\ln x$ by shifting left one, up two units, and x-axis reflection.

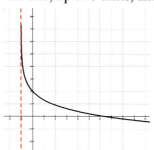

ALGEBRA REVIEW

For the following functions, find $f \circ g = f(g(x))$ and $g \circ f = g(f(x))$.

1. $f(x) = 3x - 5$ and $g(x) = x^2 + 1$

2. $f(x) = \sqrt{x}$ and $g(x) = 4x + 3$

3. $f(x) = x^2 + 2x$ and $g(x) = x - 1$

For the following functions, find the inverse function. Graph the function and its inverse. Determine the domain and range of the function and its inverse.

4. $f(x) = 2x - 1$

5. $f(x) = (x-1)^2$ for $x \geq 1$

6. $f(x) = \sqrt[3]{x} + 3$

Lesson 8 — Properties of Logarithms

Basic Properties of Logarithms

1. $\log_b 1 = 0$ Since $b^0 = 1$, $\log_b 1 = 0$.

2. $\log_b b = 1$ Since $b^1 = b$, $\log_b b = 1$.

3. $\log_b b^x = x$

4. $b^{\log_b x} = x$

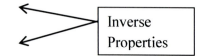

Inverse Properties

Example 1: Use the properties of logarithms to evaluate the expressions without a calculator.

  a. $\ln 1$ **b.** $\log 10$ **c.** $\log_2 64$ **d.** $e^{\ln 4}$

Solutions:

a. Since $e^0 = 1$, $\ln 1 = 0$.   *The answer to a logarithmic expression is the exponent.*

b. Since $10^1 = 10$, $\log 10 = 1$.

c. Since $2^6 = 64$, $\log_2 2^6 = 6$

d. Since e^x and $\ln x$ are inverses, $e^{\ln 4} = 4$

Product Rule for Logarithms

$$\log_b MN = \log_b M + \log_b N$$

Proof: Let $x = \log_b M$ and $y = \log_b N$. Using the exponential form, this means $b^x = M$ and $b^y = N$. Multiply M and N.

$$MN = b^x b^y$$

$$MN = b^{x+y}$$ *Rule of Exponents: When multiplying like bases, add exponents.*

Take the logarithm of base b on both sides of the equation. $\log_b MN = \log_b b^{x+y}$

Use the inverse property to simplify the right side. $\log_b MN = x + y$

Now substitute for $x = \log_b M$ and $y = \log_b N$. $\log_b MN = \log_b M + \log_b N$

Quotient Rule for Logarithms

$$\log_b \frac{M}{N} = \log_b M - \log_b N$$

Proof: Let $x = \log_b M$ and $y = \log_b N$. Using the exponential form, this means $b^x = M$ and $b^y = N$. Divide M and N.

$$\frac{M}{N} = \frac{b^x}{b^y} = b^{x-y}$$ Rule of Exponents: When dividing like bases, subtract exponents.

Take the logarithm of base b on both sides of the equation. $\log_b \frac{M}{N} = \log_b b^{x-y}$

Use the inverse property to simplify the right side. $\log_b \frac{M}{N} = x - y$

Now substitute for $x = \log_b M$ and $y = \log_b N$. $\log_b \frac{M}{N} = \log_b M - \log_b N$

Power Rule for Logarithms

$$\log_b M^p = p \log_b M$$

Proof: Let $x = \log_b M$. Using the exponential form, this means $M = b^x$.

Raise both sides of the equation to the power of p. $M^p = (b^x)^p = b^{xp}$ Multiply exponents.

Take the logarithm of base b on both sides of the equation. $\log_b M^p = \log_b b^{xp}$

Use the inverse property to simplify the right side. $\log_b M^p = xp$

Now substitute for $x = \log_b M$. $\log_b M^p = p \log_b M$

The proofs of the logarithm rules are straight from the rules of exponents. If you multiply same bases, add the exponents. If you divide same bases, subtract the exponents. If you have a power raised to another power, multiply the exponents. Since a logarithmic expression represents an exponent, it makes sense that we follow the same rules.

Example 2: Use the properties of logarithms to condense each expression as a single term.
a. $\ln x + \ln y$
b. $3\log x - 4\log y + 2\log z$
c. $3\log x - (4\log y + 2\log z)$

Solution:

a. $\ln x + \ln y = \ln(xy)$ Use the Product Rule

b. $3\log x - 4\log y + 2\log z = \log x^3 - \log y^4 + \log z^2$ 1st: Use the Power Rule

$\qquad = \log \dfrac{x^3}{y^4} + \log z^2$ 2nd: Use the Quotient Rule

$\qquad = \log \dfrac{x^3 z^2}{y^4}$ 3rd: Use the Product Rule

c. $3\log x - (4\log y + 2\log z) = \log x^3 - (\log y^4 + \log z^2)$ 1st: Use the Power Rule

$\qquad = \log x^3 - (\log y^4 z^2)$ 2nd: Use the Power Rule

$\qquad = \log \dfrac{x^3}{y^4 z^2}$ 3rd: Use the Quotient Rule

Why do we need to condense logarithms? It is important in any math class to write your answer in simplified or reduced form. The same is true for logarithms. Why write 2 or 3 logarithms when you can simplify down to one single logarithm? In upper level math classes, you will find an advantage by writing logarithms as sums and/or differences using properties of logarithms. This is called expanding the logarithms. Practice condensing and expanding logarithms, so you can apply the rules of logarithms in any situation.

Example 3: Use the properties of logarithms to expand each expression as sums or differences. Simplify if possible.

a. $\log(100x)$
b. $\log_2\left(\dfrac{8x}{y}\right)$
c. $\ln(5x^2 y^3)$
d. $\log_3 \sqrt{\dfrac{xy}{z}}$

Solution:

a. $\log(100x) = \log 100 + \log x$
$\qquad = 2 + \log x$
$10^2 = 100$

b. $\log_2\left(\dfrac{8x}{y}\right) = \log_2(8x) - \log_2 y$
$\qquad = \log_2 8 + \log_2 x - \log_2 y$
$\qquad = 3 + \log_2 x - \log_2 y$
$2^3 = 8$

c. $\ln(5x^2 y^3) = \ln 5 + \ln x^2 + \ln y^3$
$= \ln 5 + 2\ln x + 3\ln y$

d. $\log_3 \sqrt{\dfrac{xy}{z}} = \dfrac{1}{2}\log_3 \dfrac{xy}{z}$
$= \dfrac{1}{2}[\log_3(xy) - \log_3 z]$
$= \dfrac{1}{2}(\log_3 x + \log_3 y - \log_3 z)$

Change of Base Formula

$$\log_b M = \dfrac{\log_a M}{\log_a b}$$

Proof: Let $x = \log_b M$. Using the exponential form, this means $b^x = M$.

Take the logarithm of base a on both sides of the equation, where $a > 0$ and $a \neq 1$.

$$\log_a b^x = \log_a M$$

Use the power rule for logarithms. $\quad x \log_a b = \log_a M$

Solve for x by dividing both sides by $\log_a b$. $\quad x = \dfrac{\log_a M}{\log_a b}$

Substitute for x. $\quad \log_b M = \dfrac{\log_a M}{\log_a b}$

Example 4: Use the change of base formula to round the logarithmic expression to 3 decimal places.

a. $\log_4 56$

b. $\log_{1/2} 27$

Solution: According to the change of base formula, any base would work. Since we want to use the calculator, use base 10 or base e.

a. $\log_4 56 = \dfrac{\log 56}{\log 4} = 2.904$ or $\log_4 56 = \dfrac{\ln 56}{\ln 4} = 2.904$

Note: The number 2.904 is the exponent that results in 56. $4^{2.904} \approx 56$

b. $\log_{1/2} 27 = \dfrac{\log 27}{\log \dfrac{1}{2}} = -4.755$ or $\log_{1/2} 27 = \dfrac{\ln 27}{\ln \dfrac{1}{2}} = -4.755$

The number -4.755 is the exponent that results in 27. $\left(\dfrac{1}{2}\right)^{-4.755} \approx 27$

Example 5: Use the change of base formula to find the value of the logarithmic expression without using a calculator. $(\log_3 5)(\log_5 7)(\log_7 9)$

Solution: Changing the base of each logarithm to base 10, we can see that some of the logarithms will cancel when multiplied.

$$(\log_3 5)(\log_5 7)(\log_7 9) = \left(\frac{\log 5}{\log 3}\right)\left(\frac{\log 7}{\log 5}\right)\left(\frac{\log 9}{\log 7}\right)$$

$$= \left(\frac{\log 9}{\log 3}\right)$$

$$= \log_3 9$$

$$= 2$$

$3^2 = 9$

You can also use the Change of Base Formula to simplify a logarithm expression. $\dfrac{\log_a M}{\log_a b} = \log_b M$

To graph $f(x) = -7\log_4(x-1) + 3$, we could use transformations on the base function $y = \log_4 x$. The graph would shift right 1 unit, reflect the x–axis, stretch vertically by a factor of 7, and up 3 units. With the change of base formula, we can use the calculator to graph the function. Change the base to the common base 10 or the natural base e.

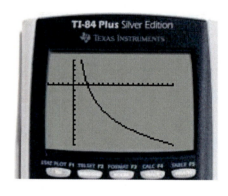

$$-7\log_4(x-1) + 3 = \frac{-7\log(x-1)}{\log(4)} + 3 \quad \text{or} \quad \frac{-7\ln(x-1)}{\ln(4)} + 3$$

Be careful not to make up your own properties or confuse the properties. Determine if the following statements are true or false for all $x > 0$ and $y > 0$.

1.) $\log_b(x + y) = \log_b x + \log_b y$
2.) $\log_b(xy) = \log_b x \cdot \log_b y$
3.) $\log_b(xy) = \log_b x + \log_b y$
4.) $\log_b x \cdot \log_b y = \log_b x + \log_b y$
5.) $\log_b x - \log_b y = \log_b(x - y)$
6.) $\log_b \dfrac{x}{y} = \dfrac{\log_b x}{\log_b y}$
7.) $\dfrac{\log_b x}{\log_b y} = \log_b x - \log_b y$
8.) $\log_b(x^n) = n\log_b x$
9.) $(\log_b x)^n = n\log_b x$
10.) $\log_b \sqrt{x} = \dfrac{1}{2}\log_b x$

These statements are created when a student doesn't have a full understanding of the logarithmic properties. Make sure you know how to apply the properties correctly. (Only numbers 3, 8, and 10 are true!)

Lesson 8 — Practice Exercises

In Exercises 1 – 10, evaluate the logarithm without the use of a calculator.

1. $\log 100$
2. $\log_2 16$
3. $\log_3 81$
4. $\log_{25} 5$
5. $\log_2 \frac{1}{8}$

6. $\log_3 3$
7. $\log_4 \frac{1}{16}$
8. $\ln e$
9. $\ln \frac{1}{e}$
10. $\ln e^2$

In Exercises 11 – 14, use the change of base formula to evaluate the logarithm.

11. $\log_2 48$
12. $\log_3 15$
13. $\log_4 125$
14. $\log_5 51$

In Exercises 15 – 28, use the properties of logarithms to condense each expression as a single term. Simplify where possible.

15. $\log_2 160 - \log_2 5$
16. $\log 50 + \log 20$
17. $\log_3 7 + \log_3 x$

18. $\log x + 2 \log y$
19. $3 \log x - \log y + 4 \log z$
20. $\log x - \frac{1}{2} \log y - \log z$

21. $2(\log x - \log y) + \log z$
22. $\frac{1}{2}(\log x + \log y) - \log z$
23. $\frac{1}{2}(\ln x + \ln y + \ln z)$

24. $\ln(x^2 - 25) - \ln(x - 5)$
25. $\log(x + 1) - \log(x^2 - 1)$
26. $\ln x + \ln(x - 2) - \ln(x + 1)$

27. $2[\ln x - \ln(x + 3)] + \frac{1}{2} \ln(x - 4)$
28. $3\log(x + 3) + 4\log(x + 1) + 2\log(x - 7)$

In Exercises 29 – 42, use the properties of logarithms to expand each expression as sums or differences. Simplify where possible.

29. $\log 100x^2$
30. $\log_2 8(x + 1)$
31. $\log_2 \frac{64}{xy}$

32. $\log \frac{x + 4}{10y}$
33. $\log \frac{10x^3}{y + 1}$
34. $\log_2 \frac{4xy}{x^3}$

35. $\ln \sqrt{\frac{x}{y}}$
36. $\log \sqrt[3]{x(x + 4)}$
37. $\log_2 \sqrt{\frac{x + 5}{x - 9}}$

38. $\ln[x(x + 5)]^2$
39. $\ln(x\sqrt{y})$
40. $\log(x^2 y^3 \sqrt{z})$

41. $\log \frac{(x + 1)^3}{x(x - 3)^2}$
42. $\log \sqrt{\frac{(x - 6)(x + 2)}{(x + 4)(x - 2)}}$

Solutions to Practice Exercises   **Lesson 8**

1. 2   2. 4   3. 4   4. $\dfrac{1}{2}$ 5. -3

6. 1   7. -2 8. 1   9. -1 10. 2

11. $\dfrac{\ln 48}{\ln 2}$ or $\dfrac{\log 48}{\log 2} \approx 5.585$ 12. $\dfrac{\ln 15}{\ln 3}$ or $\dfrac{\log 15}{\log 3} \approx 2.465$

13. $\dfrac{\ln 125}{\ln 4}$ or $\dfrac{\log 125}{\log 4} \approx 3.483$ 14. $\dfrac{\ln 51}{\ln 5}$ or $\dfrac{\log 51}{\log 5} \approx 2.443$

15. $\log_2 \dfrac{160}{5} = \log_2 32 = 5$ 16. $\log(50 \cdot 20) = \log 1000 = 3$

17. $\log_3(7x)$ 18. $\log(xy^2)$ 19. $\log \dfrac{x^3 z^4}{y}$ 20. $\log \dfrac{x}{z\sqrt{y}}$

21. $\log \dfrac{x^2 z}{y^2}$ 22. $\log \dfrac{\sqrt{xy}}{z}$ 23. $\ln \sqrt{xyz}$ 24. $\ln \dfrac{x^2 - 25}{x - 5} = \ln(x+5)$

25. $\log \dfrac{1}{x-1}$ 26. $\ln \dfrac{x(x-2)}{x+1}$ 27. $\ln \dfrac{x^2 \sqrt{x-4}}{(x+3)^2}$ 28. $\log[(x+3)^3 (x+1)^4 (x-7)^2]$

29. $\log 100 + 2\log x = 2 + 2\log x$ 30. $\log_2 8 + \log_2(x+1) = 3 + \log_2(x+1)$

31. $\log_2 64 - \log_2 x - \log_2 y = 6 - \log_2 x - \log_2 y$ 32. $\log(x+4) - \log 10 - \log y = \log(x+4) - 1 - \log y$

33. $\log 10 + 3\log x - \log(y+1) = 1 + 3\log x - \log(y+1)$

34. $\log_2 4 + \log_2 x + \log_2 y - 3\log_2 x = 2 + \log_2 y - 2\log_2 x$

35. $\dfrac{1}{2}(\ln x - \ln y)$ 36. $\dfrac{1}{3}[\log x + \log(x+4)]$ 37. $\dfrac{1}{2}[\log_2(x+5) - \log_2(x-9)]$

38. $2[\ln x + \ln(x+5)]$ 39. $\ln x + \dfrac{1}{2}\ln y$ 40. $2\log x + 3\log y + \dfrac{1}{2}\log z$

41. $3\log(x+1) - \log x - 2\log(x-3)$ 42. $\dfrac{1}{2}[\log(x-6) + \log(x+2) - \log(x+4) - \log(x-2)]$

Lesson 9   Exponential and Logarithmic Equations

To solve equations, you must use inverse operations to isolate the variable.

The inverse operation for an exponential expression is a logarithm. The inverse operation for a logarithmic expression is an exponent.

> **Steps for Solving Exponential Equations:**
> 1. Isolate the exponential expression.
> 2. Take a logarithm on both sides of the equation.
> 3. Solve for x.
> 4. Check your solution in the original equation.

Example 1: Solve the exponential equation for x.

$$2 - 3e^{0.4x} = -7$$

Solution: To solve for x, use inverse operations.

$2 - 3e^{0.4x} = -7$ Subtract 2.

$-3e^{0.4x} = -9$ Divide by -3.

$e^{0.4x} = 3$

$\ln e^{0.4x} = \ln 3$ To remove the exponential with base e, take the natural logarithm of both sides.

$0.4x = \ln 3$ Remember, $\ln e = 1$.

$x = \dfrac{\ln 3}{0.4}$ Divide by 0.4 to finish the solving process.

The exact answer is $x = \dfrac{\ln 3}{0.4}$. To get an approximation, use your calculator. Don't forget to put parentheses around 3. $x = \ln(3)/0.4 \approx 2.747$

Check: $2 - 3e^{0.4(2.747)} \approx -7$

When solving exponential equations, one technique requires rewriting the numbers with the same base. For example, 144 is the same thing as 12^2. Therefore, if the equation were $144 = 12^x$, we would replace 144 with 12^2. $12^2 = 12^x$ If the bases are equal, then the exponents are also equal.

Property: If $b^x = b^y$, then $x = y$

Example 2: Solve the following exponential equation. $\left(\dfrac{1}{9}\right)^{x-5} = 3^{3x}$

Solution: Notice, the fraction $\dfrac{1}{9}$ can be written with base 3. $\dfrac{1}{9} = \dfrac{1}{3^2} = 3^{-2}$

$\left(\dfrac{1}{9}\right)^{x-5} = 3^{3x}$ Replace $\dfrac{1}{9}$ with 3^{-2}.

$\left(3^{-2}\right)^{x-5} = 3^{3x}$

$3^{-2x+10} = 3^{3x}$ Use the power rule for exponents by multiplying $-2(x-5)$.

$-2x + 10 = 3x$ Apply the property. If the bases are equal, the exponents are equal.

$10 = 5x$

$2 = x$ Solve for x.

Check:

$\left(\dfrac{1}{9}\right)^{2-5} = 3^{3(2)}$

$\left(\dfrac{1}{9}\right)^{-3} = 3^6$

$729 = 729$

Example 3: Solve $\dfrac{e^4}{e^{2-x}} = e^3 e$.

Solution: First, notice both sides of the equation can be written with the same base e. Use the properties of exponents to combine the expressions. When dividing, exponents are subtracted. When multiplying, exponents are added.

$e^{4-(2-x)} = e^{3+1}$

$e^{2+x} = e^4$

$2 + x = 4$

$x = 2$

One application for solving exponential equation involves chemicals. The first thing you need is the rate of decay for the chemical. This can be found by solving an exponential equation. The radioactive decay is measured in terms of half-life, which is the amount of time required for half of the chemical to decay.

Example 4: **The half-life for Radium (^{226}Ra) is 1599 years. Find the decay rate.**

Solution: The exponential decay model is $y = A_0 e^{kt}$. To find the decay rate, we need to solve for k. Since the half-life is known, let $t = 1599$. A_0 represents the initial amount of Radium present. This information is not given. However, we know that $\frac{1}{2} A_0$ will be remaining after 1599 years.

$y = A_0 e^{kt}$

$\frac{1}{2} A_0 = A_0 e^{1599k}$ Divide both sides by A_0.

$\frac{1}{2} = e^{1599k}$

$\ln\left(\frac{1}{2}\right) = \ln\left(e^{1599k}\right)$ Take the natural logarithm of both sides. Using the inverse property to simplify the right side of the equation.

$\ln\left(\frac{1}{2}\right) = 1599k$ Divide by 1599.

$\dfrac{\ln\left(\frac{1}{2}\right)}{1599} = k$ The decay model for Radium can be written as $y = A_0 e^{-0.000433t}$.

-0.000433

Steps for Solving Logarithmic Equations:

1. Isolate the logarithmic expression. Condense logarithms if necessary.

2. Change the equation to exponential form.

3. Solve for x.

4. Check your solution in the original equation.

One important thing to remember is to check your solution, especially when there are domain restrictions. For the logarithmic function $f(x) = \log(x+2) - 6$, the domain would be any value greater than –2. Therefore, if you solve equations with this function, the solution would be restricted to values greater than –2.

$x + 2 > 0$

$x > -2$

Example 5: **Solve the logarithmic equation.** $\quad \dfrac{1}{2}\ln(2x+5) + 3 = 3.2$

Solution: Find the domain of the logarithm to determine the interval of possible solutions.

$2x + 5 > 0$

$x > -\dfrac{5}{2}$

The solution for this equation must be in the interval $\left(-\dfrac{5}{2}, \infty\right)$.

$\dfrac{1}{2}\ln(2x+5) + 3 = 3.2 \qquad$ Isolate the logarithm by subtracting 3 and multiplying by 2.

$\dfrac{1}{2}\ln(2x+5) = 0.2$

$\ln(2x+5) = 0.4 \qquad$ To remove the natural logarithm, use the natural base exponential.

$2x + 5 = e^{0.4} \qquad e^{\ln(2x+5)} = 2x + 5$

$2x = e^{0.4} - 5 \qquad$ Solve for x by subtracting 5.

$x = \dfrac{e^{0.4} - 5}{2} \qquad$ Divide by 2 to finish the solving process.

Use parentheses for approximating the solution in the calculator. $x = (e^{0.4} - 5)/2 \approx -1.7541$

Note: this value lies within the restricted domain of $\left(-\dfrac{5}{2}, \infty\right)$.

Example 6: **Solve the following equation.** $\quad 9 - 4\log(2x) = 3.6$

Solution: Find the domain of the logarithm to determine the interval of possible solutions.

$2x > 0$
$x > 0$

The solution for this equation must be in the interval $(0, \infty)$.

$9 - 4\log(2x) = 3.6$ Isolate the logarithmic expression by subtracting 9 and dividing by -4.
$-4\log(2x) = -5.4$
$\log(2x) = 1.35$ To remove the logarithm, use the base 10 exponential.
$2x = 10^{1.35}$ Divide by 2 to finish the solving process.
$x = \dfrac{10^{1.35}}{2}$
$x \approx 11.1936$

Property: If $\log_b x = \log_b y$, then $x = y$.

Example 7: Use properties of logarithms to solve the equation. Be sure to check the domain for restrictions.

$\log_3 x + \log_3(x+2) = \log_3 8$

Solution: Find the domain of the logarithm to determine the interval of possible solutions.

$x > 0$ and $x + 2 > 0$
$\qquad\qquad x > -2$

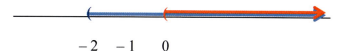

Since both restrictions must be met, we use the intersection of the inequalities. Therefore, the solution must be greater than 0.

$\log_3 x + \log_3(x+2) = \log_3 8$ Notice that the product rule can be applied to condense the
$\log_3 x(x+2) = \log_3 8$ logarithms on the left side of the equation.
$\log_3(x^2 + 2x) = \log_3 8$ $\log_b M + \log_b N = \log_b MN$
$x^2 + 2x = 8$
$x^2 + 2x - 8 = 0$ Use the above property to cancel the logarithms.
$(x+4)(x-2) = 0$ Factor the quadratic equation to solve for x.
$x + 4 = 0 \quad x - 2 = 0$
$x = -4 \quad x = 2$ Since $-4 < 0$, $x \neq -4$. The only solution is $x = 2$.

Check: $\log_3 2 + \log_3(2+2) = \log_3 8$

$\log_3(2 \cdot 4) = \log_3 8$

A value that is a solution to an intermediate equation but does not satisfy the original equation is called an **extraneous solution**. In Example 6, $x = -4$ is an extraneous solution. Don't automatically assume negative numbers will be extraneous. Some logarithms can include negative solutions due to horizontal shifts to the left.

We used logarithms to determine the level of sound intensity in Lesson 7. Logarithms can also be used to measure the magnitude of earthquakes and atmospheric pressure. In chemistry, logarithms are used to measure the *pH* level.

Example 8: H^+ is the concentration of hydrogen ions in moles per liter. The *pH* of a chemical solution is given by the formula $pH = -\log(H^+)$. Values of *pH* range from 0 to 14, with 7 considered to be neutral.

a. Determine the hydrogen ion concentration when the *pH* level is 0 (highly acidic) and 14 (highly alkaline).

b. Determine the hydrogen ion concentration of coffee and spinach.

Solution:

a. $0 = -\log(H^+)$ $14 = -\log(H^+)$

$0 = \log(H^+)$ $-14 = \log(H^+)$

$10^0 = H^+$ $10^{-14} = H^+$

$H^+ = 1$

---

b. For black coffee, $pH = 5$. $5 = -\log(H^+)$

$-5 = \log(H^+)$

$10^{-5} = H^+$

For raw spinach, $pH = 8.5$. $8.5 = -\log(H^+)$

$-8.5 = \log(H^+)$

$10^{-8.5} = H^+$

Lesson 9   Practice Exercises

In Exercises 1 – 4, write each equation in exponential form.

1. $\log_s 5 = p$
2. $\log x = y$
3. $\log_3 a = 6$
4. $\ln t = s$

In Exercises 5 – 8, write each equation in logarithmic form.

5. $r^2 = t$
6. $10^x = 6$
7. $b^a = c$
8. $e^t = d$

In Exercises 9 – 20, solve the exponential equations. Write answers in exact form and approximate the answer using three decimal places (if necessary).

9. $4^{2x} = 8^{x+1}$
10. $27^{2x+4} = 9^{4x}$
11. $e^{3x-5} = e^{x+4}$
12. $\dfrac{e^4}{e^{2-x}} = e^2 e^2$

13. $3^{x+3} = 6$
14. $2^x + 1 = 11$
15. $4(2^x) - 5 = 95$
16. $\dfrac{1}{2}(10^{x-2}) = 7.256$

17. $160 e^{0.75x} = 672$
18. $e^{-5x} + 9 = 75$
19. $3.4 e^{-x} = 176.8$
20. $45 e^{3x} + 100 = 1990$

In Exercises 21 – 30, solve the logarithmic equations. Write answers in exact form and approximate the answer using three decimal places (if necessary).

21. $\log_2(x+3) = \log_2 x + \log_2 4$
22. $\ln(x-1) + \ln 6 = \ln(3x)$
23. $\log_3(x+6) + \log_3(x+4) = 1$
24. $\log(x+6) - \log x = \log 5$
25. $\log_3 x + 10 = 7$
26. $4\ln(x-1) + 1 = 15$
27. $\log(x+3) - \log x = 1$
28. $\log(x-15) = 2 - \log x$
29. $\log_3(2x-1) - \log_3(x-4) = 2$
30. $\ln 21 = 1 + \ln(x-2)$

Solutions to Practice Exercises **Lesson 9**

1. $s^p = 5$
2. $10^y = x$
3. $3^6 = a$
4. $e^s = t$

5. $\log_r t = 2$
6. $\log 6 = x$
7. $\log_b c = a$
8. $\ln d = t$

9. 3
10. 6
11. 4.5
12. 2

13. $\dfrac{\ln 6}{\ln 3} - 3 \approx -1.369$
14. $\dfrac{\ln 10}{\ln 2} \approx 3.322$
15. $\dfrac{\ln 25}{\ln 2} \approx 4.644$
16. $\log(14.512) + 2 \approx 3.162$

17. $\dfrac{\ln 4.2}{0.75} \approx 1.913$
18. $\dfrac{\ln 66}{-5} \approx -0.838$
19. $-\ln 52 \approx -3.951$
20. $\dfrac{\ln 42}{3} \approx 1.246$

21. 1
22. 2
23. -3 (-7 is extraneous)
24. $\dfrac{3}{2}$

25. $\dfrac{1}{27}$
26. $e^{3.5} + 1 \approx 34.115$
27. $\dfrac{1}{3}$
28. 20 (-5 is extraneous)

29. 5
30. $\dfrac{21}{e} + 2 \approx 9.725$

ALGEBRA REVIEW

Population grows exponentially at first but will experience logistic growth overtime due to limited resources. There is a maximum number an environment can support, which is called the limiting size or carrying capacity. Using statistics from population data, a logistic growth model for world population is found to be $f(t) = \dfrac{11.778}{1 + 2.914e^{-0.0284t}}$, where t is the number of years after 1960 and $f(t)$ is measured in billions.

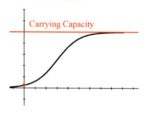
Carrying Capacity

1. According to the model, what was the world population in 1960? How accurate is the model if the population was actually 3.04 billion in 1960 according to the census bureau?

2. What will be the world population in 2020?

3. When will the population reach 10 billion?

4. What is the limiting size of the world population according to the model?

Cumulative Review 2   Lessons 5 – 9

1. Find the vertical asymptote(s), horizontal asymptote, slant/oblique asymptote for the following. Find the points (if any) where the graph crosses the asymptotes.

   a. $f(x) = \dfrac{3x^2 + 2x - 4}{2x^2 - x + 1}$
   b. $g(x) = \dfrac{x^2 - 4}{x + 1}$
   c. $h(x) = \dfrac{x - 3}{x^2 - 5x - 6}$
   d. $m(x) = \dfrac{x^2 - x - 12}{x - 4}$
   e. $f(x) = \dfrac{4 - x}{2x + 3}$
   f. $g(x) = \dfrac{8x^3}{x^2 + 4}$

2. Find the following for each function and graph.
   Domain, Hole, Vertical Asymptote, Horizontal Asymptote (does the graph cross and where), Slant/Oblique Asymptote (does the graph cross and where), x-intercept, y-intercept

   a. $f(x) = \dfrac{x - 2}{x^2 - x - 2}$
   b. $g(x) = \dfrac{2x^2 + 1}{x^2}$
   c. $h(x) = \dfrac{x^2 - 3}{x + 1}$

3. Solve and graph. Write answers in interval notation.

   a. $x^2 + 6 < -5x$
   b. $x^2 - 3x \geq 28$
   c. $\dfrac{-3}{2 - x} \leq 0$
   d. $\dfrac{x + 1}{x - 2} + \dfrac{x - 3}{x - 1} < 0$
   e. $\dfrac{x + 1}{x - 2} - 4 \geq 0$
   f. $\dfrac{3}{x^2 - 4} \leq \dfrac{5}{x^2 + 7x + 10}$

4. Graph the following and state the domain, range, and asymptote.

   a. $f(x) = 2^x + 4$
   b. $f(x) = 2^{x-1} - 3$
   c. $f(x) = e^x - 1$
   d. $f(x) = \log_2(x + 1)$
   e. $f(x) = \log_2 x - 1$
   f. $f(x) = \ln(x + 1) - 3$

5. Graph the function and its inverse using the same set of axes.

   a. $f(x) = 2^x$ and $f^{-1}(x) = \log_2 x$
   b. $f(x) = 10^x$ and $f^{-1}(x) = \log x$
   c. $f(x) = e^x$ and $f^{-1}(x) = \ln x$

6. Apply the properties of logarithms and write the logarithm as a sum and difference.

   a. $\log_a 8x^3 y^2 z$
   b. $\log_a \sqrt{\dfrac{x^2 y^3}{z^4}}$
   c. $\log_a \sqrt[3]{\dfrac{x^6 y^2}{a^4 b^3}}$

7. Express a single logarithm.

   a. $2\ln x - 3\ln y$
   b. $\dfrac{2}{3}\left[\ln(x^2 - 9) - \ln(x + 3)\right]$
   c. $\ln 54 - \ln 6$
   d. $\log 0.01 + \log 10{,}000$
   e. $\log_a x + 3\log_a y - \dfrac{1}{2}\log_a z$

8. Simplify.

   a. $\log_a a^4$
   b. $10^{\log x^4}$
   c. $e^{\ln x^3}$
   d. $\log 1000$

9. Solve the equations. Round your answers to two decimal places, if necessary.

   a. $32^{2x} = 16^{x-3}$
   b. $e^{x-1} = \left(\dfrac{1}{e^4}\right)^{x+1}$
   c. $e^{5x-3} - 2 = 10476$
   d. $\log_5 x + \log_5(4x-1) = 1$
   e. $6 + 2\ln x = 5$
   f. $\log_2(x+2) - \log_2(x-5) = 3$

Solutions to Review

1a. Vertical Asymptote: none Horizontal Asymptote: $y = \dfrac{3}{2}$, crosses at $\left(\dfrac{11}{7}, \dfrac{3}{2}\right)$

1b. Vertical Asymptote: $x = -1$ Slant/Oblique Asymptote: $y = x - 1$, doesn't cross

1c. Vertical Asymptote: $x = 6, x = -1$ Horizontal Asymptote: $y = 0$, crosses $(3,0)$

1d. NONE, but there is a hole in the graph at $(4,7)$.

1e. Vertical Asymptote: $x = -\dfrac{3}{2}$ Horizontal Asymptote: $y = -\dfrac{1}{2}$

1f. Vertical Asymptote: none Slant/Oblique Asymptote: $y = 8x$, cross at $(0,0)$

2a. Domain: $(-\infty, -1) \cup (-1, 2) \cup (2, \infty)$ Hole: $\left(2, \dfrac{1}{3}\right)$
    Vertical Asymptote: $x = -1$ Horizontal Asymptote: $y = 0$, doesn't cross
    x-intercept: none y-intercept: $(0,1)$

2b. Domain: $(-\infty, 0) \cup (0, \infty)$ Hole: No
    Vertical Asymptote: $x = 0$ Horizontal Asymptote: $y = 2$, doesn't cross
    x-intercept: none y-intercept: none

2c. Domain: $(-\infty, -1) \cup (-1, \infty)$ Hole: No
    Vertical Asymptote: $x = -1$ Slant/Oblique Asymptote: $y = x - 1$, doesn't cross
    x-intercept: $\left(\pm\sqrt{3}, 0\right)$ y-intercept: $(0, -3)$

Graphs:

2a. 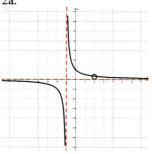 2b. 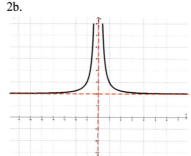 2c.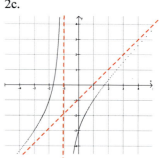

3a. $(-3, -2)$ 3b. $(-\infty, -4] \cup [7, \infty)$ 3c. $(-\infty, 2)$

3d. $(1, 2)$ 3e. $(2, 3]$ 3f. $(-\infty, -5) \cup (-2, 2) \cup [12.5, \infty)$

4a. Domain $(-\infty, \infty)$, Range $(4, \infty)$, Asymptote $y = 4$

4b. Domain $(-\infty, \infty)$, Range $(-3, \infty)$, Asymptote $y = -3$

4c. Domain $(-\infty, \infty)$, Range $(-1, \infty)$, Asymptote $y = -1$

Graphs:
4a.  4b.  4c.

4d. Domain $(-1, \infty)$, Range $(-\infty, \infty)$, Asymptote $x = -1$

4e. Domain $(0, \infty)$, Range $(-\infty, \infty)$, Asymptote $x = 0$

4f. Domain $(-1, \infty)$, Range $(-\infty, \infty)$, Asymptote $x = -1$

Graphs:
4d. 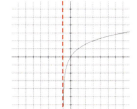 4e. 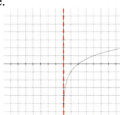 4f.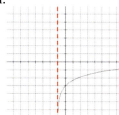

5a. 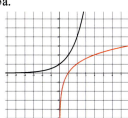 5b. 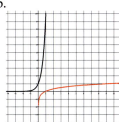 5c.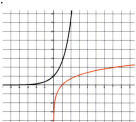

6a. $3\log_a 2 + 3\log_a x + 2\log_a y + \log_a z$ 6b. $\log_a x + \tfrac{3}{2}\log_a y - 2\log_a z$

6c. $2\log_a x + \tfrac{2}{3}\log_a y - \tfrac{4}{3} - \log_a b$

7a. $\ln \dfrac{x^2}{y^3}$ 7b. $\ln \sqrt[3]{(x-3)^2}$ 7c. $\ln 9$ 7d. 2 7e. $\log_a \dfrac{xy^3}{\sqrt{z}}$

8a. 4   8b. x^4 8c. x^3 8d. 3

9a. −2   9b. −3/5   9c. 2.45

9d. 5/4   9e. 0.61   9f. 6

Lesson 10 — Introduction to Matrices

Suppose you have three points on a graph, and you need to know the equation $y = ax^2 + bx + c$ of the parabola that plots these points. How can you find the values of $a, b,$ and c?

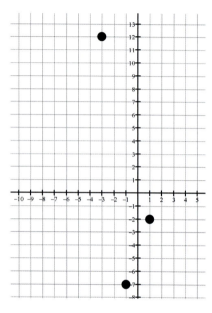

Take each point and evaluate the equation.

$(x, y) \qquad y = ax^2 + bx + c$

$(-3, 12) \qquad 12 = a(-3)^2 + b(-3) + c$

$(-1, -7) \qquad -7 = a(-1)^2 + b(-1) + c$

$(1, -2) \qquad -2 = a(1)^2 + b(1) + c$

The result is a system of equations with three variables.

$9a - 3b + c = 12$
$a - b + c = -7$
$a + b + c = -2$

If we solve the system of equations, we will find the values of each variable. This will give us the equation of the parabola that connects these three points. How do you solve the system? In this section, we will use a matrix. A **matrix** is a rectangular array of numbers. Matrices are used to organize numbers in various applications such as computer programming and cryptography. Matrices are also used on travel websites to find a flight for a certain date and price.

A line is drawn to represent the location of the equal sign.

Omitting the variables in the system of equations, the resulting matrix is $\begin{bmatrix} 9 & -3 & 1 & | & 12 \\ 1 & -1 & 1 & | & -7 \\ 1 & 1 & 1 & | & -2 \end{bmatrix}$. This matrix is called an **augmented matrix** because it contains the coefficients for the variables and also the constant terms. There are 3 **rows** (one for each equation) and 4 **columns** (three variables and the constant terms). The **order of a matrix** is $m \times n$, where m is the number of rows (horizontal entries) and n is the number of columns (the vertical entries). Therefore, the order of this matrix is 3×4. A matrix is normally named using a capital letter. We will name our matrix above, $A = \begin{bmatrix} 9 & -3 & 1 & | & 12 \\ 1 & -1 & 1 & | & -7 \\ 1 & 1 & 1 & | & -2 \end{bmatrix}$. The **entries**, a_{ij}, are the numbers in a matrix. We can label an entry using the ith row and j^{th} column. For example, entry a_{12} is the number in row 1 and column 2. Entry a_{24} is the number in row 2 and column 4. Note, $a_{12} = -3$ and $a_{24} = -7$. What is a_{33} for matrix A?

Entries a_{11}, a_{22}, a_{33} lie on the main diagonal of a matrix. In general, entry a_{ii} lies on the main diagonal of a matrix. (the row number and column number are the same)

Back to our problem: we need to find the values of $a, b,$ and c to determine the parabola that plots the three points. Using a matrix, we can perform row operations to transform the matrix into **row-echelon form**. There are only 3 row operations. This process is called **Gaussian elimination**.

Row-Equivalent Operations

1. Interchange any two rows.

2. Multiply each entry in a row by the same nonzero constant.

3. Add a nonzero multiple of one row to another row.

For a matrix to be in row-echelon form, we need to get entry a_{11} equal to 1.

$$A = \begin{bmatrix} 9 & -3 & 1 & | & 12 \\ 1 & -1 & 1 & | & -7 \\ 1 & 1 & 1 & | & -2 \end{bmatrix}$$

> **Don't forget!**
>
> 1. Rows represent equations. If we were solving this system in an algebra class, rewriting the rows in a different order does not change the system.
>
> 2. If we multiply a row by a nonzero constant, this does not change the value of the equation. Take equation 1 and multiply everything by -4
>
> $$9a - 3b + c = 12$$
> $$-36a + 12b - 4c = -48$$
>
> The equation is still balanced.
>
> 3. If we multiply equation 2 by -9 and add it to equation 1, we will cancel out the variable a.
>
> $$9a - 3b + c = 12$$
> $$\underline{-9(a - b + c = -7)}$$
> $$6b - 8c = 75$$

This can be done by interchanging row 1 and row 2.

$$A = \begin{bmatrix} 1 & -1 & 1 & | & -7 \\ 9 & -3 & 1 & | & 12 \\ 1 & 1 & 1 & | & -2 \end{bmatrix}$$

The next step is to get the entries below the main diagonal position equal to zero. In other words, we need a_{21} and a_{31} to equal zero. This can be accomplished using the 3rd row operation.

Multiply row 1 by -9, then add to row 2.

$$\begin{array}{rrrrl} -9(1 & -1 & 1 & -7) = -9 & 9 & -9 & 63 & -9R1 + R2 \to R2 \\ & & & 9 & -3 & 1 & 12 \\ \hline & & & 0 & 6 & -8 & 75 & \text{New row 2} \end{array}$$

Multiply row 1 by -1, then add to row 3.

$$\begin{array}{rrrrl} -1(1 & -1 & 1 & -7) = -1 & 1 & -1 & 7 & -R1 + R3 \to R3 \\ & & & 1 & 1 & 1 & -2 \\ \hline & & & 0 & 2 & 0 & 5 & \text{New row 3} \end{array}$$

The result is $A = \begin{bmatrix} 1 & -1 & 1 & | & -7 \\ 0 & 6 & -8 & | & 75 \\ 0 & 2 & 0 & | & 5 \end{bmatrix}$. This is the beginning of Gaussian elimination, and we continue this process until the matrix is in row-echelon form.

Row-Echelon Form (REF)

1. If a row does not consist entirely of 0's, then the first nonzero element in the row is a 1.

2. For any two successive nonzero rows, the leading 1 in the lower row is farther to the right than the leading 1 in the higher row.

3. All the rows consisting of entirely of 0's are at the bottom of the matrix.

To get matrix A into REF, we move to column two. We must get a leading 1 in row 2. This can be done by using the 2nd row operation. Multiply row 2 by the $\frac{1}{6}$.

$$A = \begin{bmatrix} 1 & -1 & 1 & | & -7 \\ 0 & 6 & -8 & | & 75 \\ 0 & 2 & 0 & | & 5 \end{bmatrix} \quad \frac{1}{6}R2 \to R2 \quad A = \begin{bmatrix} 1 & -1 & 1 & | & -7 \\ 0 & 1 & -\frac{4}{3} & | & \frac{25}{2} \\ 0 & 2 & 0 & | & 5 \end{bmatrix}$$

The next step is to get the entries beneath the main diagonal position equal to zero. In other words, we need $a_{32} = 0$. This can be accomplished using the 3rd row operation. Multiply row 2 by -2, then add to row 3.

$$-2\left(0 \quad 1 \quad -\frac{4}{3} \quad \frac{25}{2}\right) = 0 \quad -2 \quad \frac{8}{3} \quad -25 \quad\quad -2R2 + R3 \to R3$$

$$\underline{0 \quad 2 \quad 0 \quad\quad 5}$$

$$0 \quad 0 \quad \frac{8}{3} \quad -20 \quad \text{New row 3} \quad\longrightarrow\quad A = \begin{bmatrix} 1 & -1 & 1 & | & -7 \\ 0 & 1 & -\frac{4}{3} & | & \frac{25}{2} \\ 0 & 0 & \frac{8}{3} & | & -20 \end{bmatrix}$$

Notice, we have 1's on the main diagonal position for row 1 and 2. Our last step is to perform row operations to get position $a_{33} = 1$. This can be accomplished using the 2nd row operation. Multiply row 3 by $\frac{3}{8}$.

$$\frac{3}{8}\left(0 \quad 0 \quad \frac{8}{3} \quad -20\right) = 0 \quad 0 \quad 1 \quad -\frac{15}{2} \quad \text{New row 3}$$

$$A = \begin{bmatrix} 1 & -1 & 1 & | & -7 \\ 0 & 1 & -\dfrac{4}{3} & | & \dfrac{25}{2} \\ 0 & 0 & 1 & | & -\dfrac{15}{2} \end{bmatrix}$$

By Gaussian elimination, our resulting matrix is in row-echelon form. From this matrix, we can see that

$$a - b + c = -7$$
$$b - \dfrac{4}{3}c = \dfrac{25}{2}$$
$$c = -\dfrac{15}{2}$$

Using back substitution, we can find a and b.

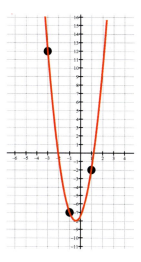

$$b - \dfrac{4}{3}\left(-\dfrac{15}{2}\right) = \dfrac{25}{2} \qquad a - \left(\dfrac{5}{2}\right) + \left(-\dfrac{15}{2}\right) = -7$$
$$b = \dfrac{5}{2} \qquad\qquad\qquad a = 3$$

Now that we know the values of $a, b,$ and c, we can find the parabola in general form.

$$y = ax^2 + bx + c \rightarrow y = 3x^2 + \dfrac{5}{2}x - \dfrac{15}{2}$$

Example 1: Solve the system of equations using Gaussian elimination.

$$\begin{cases} 2x + y = 1 \\ 3x + 4y = 14 \end{cases}$$

Solution: First, write the system as an augmented matrix.

$$\begin{bmatrix} 2 & 1 & | & 1 \\ 3 & 4 & | & 14 \end{bmatrix}$$

Our goal is to write the matrix in row-echelon form, which means we need 1's on the main diagonal and 0's under the main diagonal. Start with column 1. We need $a_{11} = 1$. It is very common to interchange two rows, but in this problem it would not help. So, we need to look at another row operation. There are several options, but we will use the 2nd row operation. Multiply row 1 by $\frac{1}{2}$.

$\begin{bmatrix} 2 & 1 & | & 1 \\ 3 & 4 & | & 14 \end{bmatrix} \; \frac{1}{2}R1 \to R1$ ← It is important to note the row operation performed on the matrix at each step.

$\begin{bmatrix} 1 & \frac{1}{2} & | & \frac{1}{2} \\ 3 & 4 & | & 14 \end{bmatrix} \; -3R1 + R2 \to R2$

After getting the leading 1 in the main diagonal, you need to get a 0 in entry a_{21}. Use the 3rd row operation. Multiply row 1 by -3 and add it to row 2.

$\begin{bmatrix} 1 & \frac{1}{2} & | & \frac{1}{2} \\ 0 & \frac{5}{2} & | & \frac{25}{2} \end{bmatrix} \; \frac{2}{5}R2 \to R2$

Move to column 2. You need a leading 1 in the main diagonal entry a_{22}. Multiply row 2 by $\frac{2}{5}$.

The matrix is now in row-echelon form.

To finish solving the system, we need to use back substitution to find the value of the variables. From the REF matrix, we have

$x + \frac{1}{2}y = \frac{1}{2}$
$y = 5$

Substitute 5 in for y, and you find that $x = -2$.

$x + \frac{1}{2}(5) = \frac{1}{2}$
$x = -2$

The solution is $(-2, 5)$. To check your solution, evaluate both equations in the system using the solution point. Both equations should result in a true statement.

$2(-2) + 5 = 1$ ✓
$3(-2) + 4(5) = 14$ ✓

If we continue to apply row operations until we have a matrix with 1's on the main diagonal and 0's above and below, the result is a matrix in **reduced row-echelon form**. This method is named for Carl Friedrich Gauss and Wilhelm Jordan. This method eliminates the need for back substitution.

In Example 1, to apply Gauss-Jordan elimination to solve the system, we only need one more step. Multiply row 2 by $-\frac{1}{2}$ and add it row 1.

$$\begin{bmatrix} 1 & \frac{1}{2} & | & \frac{1}{2} \\ 0 & 1 & | & 5 \end{bmatrix} \quad -\frac{1}{2}R2 + R1 \rightarrow R1$$

$$\begin{bmatrix} 1 & 0 & | & -2 \\ 0 & 1 & | & 5 \end{bmatrix}$$ The solution is easily seen to be $x = -2$ and $y = 5$ without any back substitution.

Always write your solution as an ordered pair, $(-2, 5)$.

Example 2: Solve the system of equations using Gauss-Jordan elimination.

$$\begin{cases} 3x + y - z = -4 \\ x + y - z = -2 \\ 2x - 2y + 3z = 3 \end{cases}$$

Solution: First, write the system as an augmented matrix.

$$\begin{bmatrix} 3 & 1 & -1 & | & -4 \\ 1 & 1 & -1 & | & -2 \\ 2 & -2 & 3 & | & 3 \end{bmatrix} \quad Rowswap\ R1 \leftrightarrow R2$$

To get a 1 on the main diagonal, swap rows 1 and 2.

$$\begin{bmatrix} 1 & 1 & -1 & | & -2 \\ 3 & 1 & -1 & | & -4 \\ 2 & -2 & 3 & | & 3 \end{bmatrix} \quad \begin{matrix} -3R1 + R2 \rightarrow R2 \\ -2R1 + R3 \rightarrow R3 \end{matrix}$$

Now, we need 0s below. Multiply row 1 by -3 and add row 2. This results in the new row 2.

Multiply row 1 by -2 and add row 3. This results in the new row 3.

$$\begin{bmatrix} 1 & 1 & -1 & | & -2 \\ 0 & -2 & 2 & | & 2 \\ 0 & -4 & 5 & | & 7 \end{bmatrix} \quad -\frac{1}{2}R2 \rightarrow R2$$

Move to column 2. Multiply each entry in row 2 by -1/2.

$$\begin{bmatrix} 1 & 1 & -1 & | & -2 \\ 0 & 1 & -1 & | & -1 \\ 0 & -4 & 5 & | & 7 \end{bmatrix} \quad \begin{matrix} -R2 + R1 \rightarrow R1 \\ 4R2 + R3 \rightarrow R3 \end{matrix}$$

Now, we need 0's above and below. Multiply row 2 by -1 and add row 1. This results in the new row 1.

Multiply row 2 by 4 and add row 3. This results in the new row 3.

$$\begin{bmatrix} 1 & 0 & 0 & | & -1 \\ 0 & 1 & -1 & | & -1 \\ 0 & 0 & 1 & | & 3 \end{bmatrix} \quad R3 + R2 \rightarrow R2$$

We have 1 in the main diagonal position a_{33}. We only need to fix row 2. We need a 0 in position a_{23}. Add row 3 and row 2. The result is the new row 2.

$$\begin{bmatrix} 1 & 0 & 0 & | & -1 \\ 0 & 1 & 0 & | & 2 \\ 0 & 0 & 1 & | & 3 \end{bmatrix}$$ From reduced row-echelon form, the solution is $(-1, 2, 3)$.

Some systems are inconsistent, which means they have no solution. During the process of Gaussian elimination or Gauss-Jordan elimination, you may encounter a matrix that results in a last row of 0's except for the last entry. If this happens, the system is inconsistent and does not have a solution.

Example 3: Solve the system of equations using Gauss-Jordan elimination.
$$\begin{cases} x+2y-2z=1 \\ 2x+2y-z=6 \\ -3x-4y+3z=-5 \end{cases}$$

Solution: We have a 1 in the first main diagonal position, so we can move on to getting 0's below.

$$\begin{bmatrix} 1 & 2 & -2 & | & 1 \\ 2 & 2 & -1 & | & 6 \\ -3 & -4 & 3 & | & -5 \end{bmatrix} \begin{matrix} \\ -2R1+R2 \to R2 \\ 3R1+R3 \to R3 \end{matrix}$$

$$\begin{bmatrix} 1 & 2 & -2 & | & 1 \\ 0 & -2 & 3 & | & 4 \\ 0 & 2 & -3 & | & -2 \end{bmatrix} \begin{matrix} R2+R1 \to R1 \\ \\ R2+R3 \to R3 \end{matrix}$$

Notice the numbers in column 2. Normally, we would get a 1 in the main diagonal position a_{22} at this point. However, sometimes you may notice that it would be easier to get the 0's above and/or below the main diagonal first.

$$\begin{bmatrix} 1 & 0 & 1 & | & 5 \\ 0 & -2 & 3 & | & 4 \\ 0 & 0 & 0 & | & 2 \end{bmatrix} \quad -\frac{1}{2}R2 \to R2$$

$$\begin{bmatrix} 1 & 0 & 1 & | & 5 \\ 0 & 1 & -\frac{3}{2} & | & -2 \\ 0 & 0 & 0 & | & 2 \end{bmatrix}$$

Since the last row consists of 0's except for the final entry, the result is $0x+0y+0z=2$. This is a false statement. Therefore, the system is inconsistent and does not have a solution.

This system has no solution.

Some systems are dependent, which means they have many solutions. During the process of Gaussian elimination or Gauss-Jordan elimination, you may encounter a matrix that results in a last row of 0's. If this happens, the system is dependent and has many solutions. You will write the solution using a dependent variable.

Example 4: Solve the system of equations using Gauss-Jordan elimination.
$$\begin{cases} 2x-4y-3z=3 \\ x+3y+z=-1 \\ 5x+y-2z=2 \end{cases}$$

Solution: First, write the system as an augmented matrix.

$$\begin{bmatrix} 2 & -4 & -3 & | & 3 \\ 1 & 3 & 1 & | & -1 \\ 5 & 1 & -2 & | & 2 \end{bmatrix} \quad \text{Rowswap } R1 \leftrightarrow R2$$

$$\begin{bmatrix} 1 & 3 & 1 & | & -1 \\ 2 & -4 & -3 & | & 3 \\ 5 & 1 & -2 & | & 2 \end{bmatrix} \quad \begin{array}{l} -2R1 + R2 \to R2 \\ -5R1 + R3 \to R3 \end{array}$$

$$\begin{bmatrix} 1 & 3 & 1 & | & -1 \\ 0 & -10 & -5 & | & 5 \\ 0 & -14 & -7 & | & 7 \end{bmatrix} \quad -\frac{1}{10} R2 \to R2$$

$$\begin{bmatrix} 1 & 3 & 1 & | & -1 \\ 0 & 1 & \frac{1}{2} & | & -\frac{1}{2} \\ 0 & -14 & -7 & | & 7 \end{bmatrix} \quad \begin{array}{l} -3R2 + R1 \to R1 \\ 14R2 + R3 \to R3 \end{array}$$

$$\begin{bmatrix} 1 & 0 & -\frac{1}{2} & | & \frac{1}{2} \\ 0 & 1 & \frac{1}{2} & | & \frac{1}{2} \\ 0 & 0 & 0 & | & 0 \end{bmatrix}$$

Since the last row consists of 0's and the final entry is also 0, the result is 0 = 0. This is a true statement. Therefore, the system is dependent and has an infinite number of solutions.

The 1st row results in the equation:

$$x + 0y - \frac{1}{2}z = \frac{1}{2}$$

$$x = \frac{1}{2}z + \frac{1}{2}$$

The 2nd row results in the equation:

$$0x + y + \frac{1}{2}z = \frac{1}{2}$$

$$y = -\frac{1}{2}z + \frac{1}{2}$$

The solution is written as an ordered triple, using the variable z. $\left(\frac{1}{2}z + \frac{1}{2}, -\frac{1}{2}z + \frac{1}{2}, z\right)$

There are infinitely many solutions. Can you name one? Choose any value for z, and substitute into the ordered triple. For example, let $z = 0$ and the result is $\left(\frac{1}{2}, \frac{1}{2}, 0\right)$. Let $z = 2$ and the result is $\left(\frac{3}{2}, -\frac{1}{2}, 2\right)$. Since z can be any real number, there are an infinite number of ordered triples that would make the system true. Therefore, the system is dependent.

Lesson 10 — Practice Exercises

In Exercises 1 and 2, determine the order of the matrix. Then find the indicated entry.

1. $A = \begin{bmatrix} 3 & -1 & 6 & 0 & -1 \\ 7 & 1 & 0 & -11 & -5 \end{bmatrix}$

   Find a_{24} and a_{13}.

2. $B = \begin{bmatrix} 4 & -10 & -3 \\ 0 & -5 & 5 \\ 1 & -2 & 1 \end{bmatrix}$

   Find b_{21} and b_{33}.

In Exercises 3 and 4, write the augmented matrix for the system of equations.

3. $\begin{cases} 3x + 2y = 1 \\ 4x - y = -6 \end{cases}$

4. $\begin{cases} x + 2y - z = 3 \\ 3x - y + 3z = 1 \\ 2x + y - 4z = 0 \end{cases}$

In Exercises 5 and 6, write a system of equations that corresponds to the augmented matrix.

5. $\begin{bmatrix} 2 & -3 & 0 \\ 1 & 2 & 7 \end{bmatrix}$

6. $\begin{bmatrix} 1 & 1 & 1 & 1 \\ 2 & 1 & -1 & 4 \\ 1 & -1 & -3 & 2 \end{bmatrix}$

In Exercises 7 – 15, solve the system of equations using Gaussian elimination.

7. $\begin{cases} 3x - y = 6 \\ 6x + 5y = -23 \end{cases}$

8. $\begin{cases} x + 2y = 6 \\ 2x - y = -8 \end{cases}$

9. $\begin{cases} 3x - 4y = -12 \\ -6x + 8y = 24 \end{cases}$

10. $\begin{cases} x + y - z = 0 \\ 3x - y + 3z = -2 \\ x + 2y - 3z = -1 \end{cases}$

11. $\begin{cases} 4x + 2y - 3z = 4 \\ 3x + 3y - z = 4 \\ x - 2y + 3z = 11 \end{cases}$

12. $\begin{cases} x + y - z = 5 \\ x + 2y - 3z = 9 \\ x - y + 3z = 3 \end{cases}$

13. $\begin{cases} 2x + y - z = 1 \\ -3x - 3y + 2z = 1 \\ -10x - 14y + 8z = 10 \end{cases}$

14. $\begin{cases} 4x + 5y + z = 1 \\ x + y - 3z = 4 \\ 2x + 3y + 7z = -7 \end{cases}$

15. $\begin{cases} -2w + 2x + 2y - 2z = -10 \\ w + x + y + z = -5 \\ 3w + x - y + 4z = -2 \\ w + 3x - 2y + 2z = -6 \end{cases}$

In Exercises 16 – 24, solve the system of equations using Gauss-Jordan elimination.

16. $\begin{cases} 2x + y = 1 \\ 3x + 2y = -2 \end{cases}$

17. $\begin{cases} 6x + 2y = -10 \\ -3x - y = 6 \end{cases}$

18. $\begin{cases} 5x - 2y = -3 \\ 2x + 5y = -24 \end{cases}$

19. $\begin{cases} 2x - 3y + 2z = 2 \\ x + 4y - z = 9 \\ -3x + y - 5z = 5 \end{cases}$ 20. $\begin{cases} 3x + 2y + 2z = 3 \\ 2x + 2y - z = 5 \\ x - 4y + z = 0 \end{cases}$ 21. $\begin{cases} 2x - 3y + 6z = 5 \\ x - y - 2z = 2 \\ 3x - 4y + 4z = 7 \end{cases}$

22. $\begin{cases} 2x - y + z = 4 \\ 8x + y + z = 2 \\ x + 2y - z = -8 \end{cases}$ 23. $\begin{cases} x - 2y + 3z = 11 \\ 4x + 2y - 3z = 4 \\ 3x + 3y - z = 4 \end{cases}$ 24. $\begin{cases} w + x + y + z = 2 \\ w + 2x + 2y + 4z = 1 \\ -w + x - y - z = -6 \\ -w + 3x + y - z = -2 \end{cases}$

25. Use the given three points to determine the constants $a, b,$ and c in the general form of the quadratic equation, $y = ax^2 + bx + c$. Graph the parabola to check your answer.

a. $(2,6); (-1,9); (1,3)$ b. $(-1,-4); (1,6); (3,0)$

Solutions to Practice Exercises   Lesson 10

1. 2×5; $a_{24} = -11$; $a_{13} = 6$ 2. 3×3; $b_{21} = 0$; $b_{33} = 1$

3. $\begin{bmatrix} 3 & 2 & | & 1 \\ 4 & -1 & | & -6 \end{bmatrix}$ 4. $\begin{bmatrix} 1 & 2 & -1 & | & 3 \\ 3 & -1 & 3 & | & 1 \\ 2 & 1 & -4 & | & 0 \end{bmatrix}$ 5. $\begin{array}{l} 2x - 3y = 0 \\ x + 2y = 7 \end{array}$ 6. $\begin{array}{l} x + y + z = 1 \\ 2x + y - z = 4 \\ x - y - 3z = 2 \end{array}$

7. $\left(\dfrac{1}{3}, -5\right)$ 8. $(-2, 4)$ 9. $\left(\dfrac{4}{3}y - 4, y\right)$ 10. $(-2, 5, 3)$ 11. $(3, -1, 2)$

12. No solution   13. $\left(\dfrac{1}{3}z + \dfrac{4}{3}, \dfrac{1}{3}z - \dfrac{5}{3}, z\right)$ 14. $(16z + 19, -13z - 15, z)$

15. $(1, -3, -2, -1)$ 16. $(4, -7)$ 17. No solution   18. $\left(-\dfrac{63}{29}, -\dfrac{114}{29}\right)$

19. $(5, 0, -4)$ 20. $\left(\dfrac{7}{4}, \dfrac{1}{8}, -\dfrac{5}{4}\right)$ 21. $(12z + 1, 10z - 1, z)$ 22. No solution

23. $(3, -1, 2)$ 24. $(1, -2, 4, -1)$

25a. $y = 2x^2 - 3x + 4$    25b. $y = -2x^2 + 5x + 3$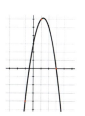

Lesson 11 — Algebra of Matrices

Two matrices are said to be equivalent if they have the same order $m \times n$ and corresponding entries are equal.

Example 1: Find the value of the missing variables to make the matrices equivalent.

$$\begin{bmatrix} 2p+1 & -5 & 9 \\ 1 & 12 & 0 \\ s+5 & 9 & -8 \end{bmatrix} = \begin{bmatrix} 7 & -5 & 2-q \\ 1 & 3r & 0 \\ -2 & 3t & -8 \end{bmatrix}$$

Solution: First, notice the order of each matrix is 3×3. For the matrices to be equivalent, the corresponding entries must be equal.

$a_{11} = b_{11}$ (1st row, 1st column) $\quad a_{12} = b_{12}$ (1st row, 2nd column) $\quad a_{13} = b_{13}$ (1st row, 3rd column)

$2p+1 = 7$ $\qquad\qquad -5 = -5 \qquad\qquad 9 = 2-q$
$\quad p = 3$ $\qquad\qquad\qquad\qquad\qquad\qquad\qquad q = -7$

$a_{21} = b_{21}$ (2nd row, 1st column) $\quad a_{22} = b_{22}$ (2nd row, 2nd column) $\quad a_{23} = b_{23}$ (2nd row, 3rd column)

$1 = 1$ $\qquad\qquad\qquad 12 = 3r \qquad\qquad\qquad 0 = 0$
$\qquad\qquad\qquad\qquad\quad r = 4$

$a_{31} = b_{31}$ (3rd row, 1st column) $\quad a_{32} = b_{32}$ (3rd row, 2nd column) $\quad a_{33} = b_{33}$ (3rd row, 3rd column)

$s+5 = -2$ $\qquad\qquad\quad 9 = 3t \qquad\qquad\qquad -8 = -8$
$\quad s = -7$ $\qquad\qquad\qquad t = 3$

In Lesson 12, we will solve matrix equations. Our result will end with two equivalent column matrices. For example,

$$\begin{bmatrix} x \\ y \\ z \end{bmatrix} = \begin{bmatrix} 1 \\ -4 \\ 7 \end{bmatrix}$$

Notice, both matrices have order 3×1. The corresponding entries would result in $x = 1$, $y = -4$, and $z = 7$.

Given two $m \times n$ matrices A and B, $A \pm B = [a_{ij} \pm b_{ij}]$. Matrix addition and subtraction is defined as adding/subtracting corresponding entries in each matrix where i is the row number and j is the column number. You cannot add or subtract matrices if the orders are not equal.

Example 2: Add or subtract the following matrices.

a. $\begin{bmatrix} 2 & -2 & 3 \\ 1 & 3 & -4 \end{bmatrix} + \begin{bmatrix} 5 & -2 & 6 \\ -2 & 3 & 5 \end{bmatrix}$

b. $\begin{bmatrix} 2 & -3 \\ -1 & 2 \\ 2 & 4 \end{bmatrix} - \begin{bmatrix} -1 & 2 \\ -4 & 1 \\ 3 & -2 \end{bmatrix}$

c. $\begin{bmatrix} 2 & -3 \\ -1 & 2 \\ 2 & 4 \end{bmatrix} - \begin{bmatrix} 2 & -2 & 3 \\ 1 & 3 & -4 \end{bmatrix}$

Solution:

a. The order of both matrices is 2×3, so these can be added. The result will be a 2×3 matrix.

$\begin{bmatrix} 2 & -2 & 3 \\ 1 & 3 & -4 \end{bmatrix} + \begin{bmatrix} 5 & -2 & 6 \\ -2 & 3 & 5 \end{bmatrix} = \begin{bmatrix} 2+5 & -2+(-2) & 3+6 \\ 1+(-2) & 3+3 & -4+5 \end{bmatrix}$

$= \begin{bmatrix} 7 & -4 & 9 \\ -1 & 6 & 1 \end{bmatrix}$

b. The order of both matrices is 3×2, so these can be subtracted. The result will be a 3×2 matrix.

$\begin{bmatrix} 2 & -3 \\ -1 & 2 \\ 2 & 4 \end{bmatrix} - \begin{bmatrix} -1 & 2 \\ -4 & 1 \\ 3 & -2 \end{bmatrix} = \begin{bmatrix} 2-(-1) & -3-2 \\ -1-(-4) & 2-1 \\ 2-3 & 4-(-2) \end{bmatrix}$

$= \begin{bmatrix} 3 & -5 \\ 3 & 1 \\ -1 & 6 \end{bmatrix}$

c. The order of the first matrix is 3×2. The order of the second matrix is 2×3. These matrices cannot be subtracted.

Normal addition is commutative. In other words, the order in which you add two numbers is not important. What about matrix addition? Is matrix addition commutative? Look at Example 2a.

$\begin{bmatrix} 2 & -2 & 3 \\ 1 & 3 & -4 \end{bmatrix} + \begin{bmatrix} 5 & -2 & 6 \\ -2 & 3 & 5 \end{bmatrix} = \begin{bmatrix} 7 & -4 & 9 \\ -1 & 6 & 1 \end{bmatrix}$

If we switch the order of the matrices, will we get the same result?

$$\begin{bmatrix} 5 & -2 & 6 \\ -2 & 3 & 5 \end{bmatrix} + \begin{bmatrix} 2 & -2 & 3 \\ 1 & 3 & -4 \end{bmatrix} = \begin{bmatrix} 5+2 & -2+(-2) & 6+3 \\ -2+1 & 3+3 & 5+(-4) \end{bmatrix} = \begin{bmatrix} 7 & -4 & 9 \\ -1 & 6 & 1 \end{bmatrix}$$

Yes! Matrix addition is commutative. For any two matrices that have the same order, $A + B = B + A$.

However, the order in which you subtract two matrices is important.

$$\begin{bmatrix} 2 & -2 & 3 \\ 1 & 3 & -4 \end{bmatrix} - \begin{bmatrix} 5 & -2 & 6 \\ -2 & 3 & 5 \end{bmatrix} \neq \begin{bmatrix} 5 & -2 & 6 \\ -2 & 3 & 5 \end{bmatrix} - \begin{bmatrix} 2 & -2 & 3 \\ 1 & 3 & -4 \end{bmatrix}$$

$$\begin{bmatrix} 2-5 & -2-(-2) & 3-6 \\ 1-(-2) & 3-3 & -4-5 \end{bmatrix} \neq \begin{bmatrix} 5-2 & -2-(-2) & 6-3 \\ -2-1 & 3-3 & 5-(-4) \end{bmatrix}$$

$$\begin{bmatrix} -3 & 0 & -3 \\ 3 & 0 & -9 \end{bmatrix} \neq \begin{bmatrix} 3 & 0 & 3 \\ -3 & 0 & 9 \end{bmatrix}$$

Matrix subtraction is not commutative. $A - B \neq B - A$

Additive Inverse (Opposite Matrix) $-A$

The Additive Inverse of a matrix can be found by replacing each entry with its opposite.

Example 3: Find the additive inverse of the following matrix A.

$$A = \begin{bmatrix} -3 & -6 \\ 4 & 3 \end{bmatrix}$$

Solution:

$$-A = \begin{bmatrix} -(-3) & -(-6) \\ -4 & -3 \end{bmatrix} = \begin{bmatrix} 3 & 6 \\ -4 & -3 \end{bmatrix}$$

What happens if you add the matrix and its additive inverse? $A + (-A)$

$$\begin{bmatrix} -3 & -6 \\ 4 & 3 \end{bmatrix} + \begin{bmatrix} 3 & 6 \\ -4 & -3 \end{bmatrix} = \begin{bmatrix} -3+3 & -6+6 \\ 4+(-4) & 3+(-3) \end{bmatrix}$$
$$= \begin{bmatrix} 0 & 0 \\ 0 & 0 \end{bmatrix}$$

Additive Identity (Zero Matrix)

The Additive Identity is a matrix consisting of only zero entries.

Scalar Multiplication

Let k be any nonzero constant. $kA = [ka_{ij}]$ is found by multiplying each entry by k.

Example 4: For matrix $A = \begin{bmatrix} -3 & -6 \\ 4 & 3 \end{bmatrix}$, find $-2A$ and $\frac{1}{3}A$.

Solution:

$$-2A = \begin{bmatrix} -2(-3) & -2(-6) \\ -2(4) & -2(3) \end{bmatrix} = \begin{bmatrix} 6 & 12 \\ -8 & -6 \end{bmatrix}$$

$$\frac{1}{3}A = \begin{bmatrix} \frac{1}{3}(-3) & \frac{1}{3}(-6) \\ \frac{1}{3}(4) & \frac{1}{3}(3) \end{bmatrix} = \begin{bmatrix} -1 & -2 \\ \frac{4}{3} & 1 \end{bmatrix}$$

Scalar multiplication can be combined with addition and subtraction. Be sure to follow the order of operations. PEMDAS…

Example 5: Perform the following operations given matrices C and D.

$$C = \begin{bmatrix} -2 & 3 \\ 4 & -2 \\ 0 & 4 \end{bmatrix} \qquad D = \begin{bmatrix} 8 & -2 \\ -3 & 2 \\ -4 & 7 \end{bmatrix}$$

a. $2C + 5D$ **b.** $-3C - 2D$ **c.** $\frac{1}{2}(C+D)$

Solution:

a. $2C + 5D$

$$\begin{bmatrix} 2(-2) & 2(3) \\ 2(4) & 2(-2) \\ 2(0) & 2(4) \end{bmatrix} + \begin{bmatrix} 5(8) & 5(-2) \\ 5(-3) & 5(2) \\ 5(-4) & 5(7) \end{bmatrix}$$

$$\begin{bmatrix} -4 & 6 \\ 8 & -4 \\ 0 & 8 \end{bmatrix} + \begin{bmatrix} 40 & -10 \\ -15 & 10 \\ -20 & 35 \end{bmatrix}$$

$$\begin{bmatrix} -4+40 & 6+(-10) \\ 8+(-15) & -4+(10) \\ 0+(-20) & 8+35 \end{bmatrix}$$

$$\begin{bmatrix} 36 & -4 \\ -7 & 6 \\ -20 & 43 \end{bmatrix}$$

b. $-3C - 2D$

$$\begin{bmatrix} -3(-2) & -3(3) \\ -3(4) & -3(-2) \\ -3(0) & -3(4) \end{bmatrix} - \begin{bmatrix} 2(8) & 2(-2) \\ 2(-3) & 2(2) \\ 2(-4) & 2(7) \end{bmatrix}$$

$$\begin{bmatrix} 6 & -9 \\ -12 & 6 \\ 0 & -12 \end{bmatrix} - \begin{bmatrix} 16 & -4 \\ -6 & 4 \\ -8 & 14 \end{bmatrix}$$

$$\begin{bmatrix} 6-16 & -9-(-4) \\ -12-(-6) & 6-4 \\ 0-(-8) & -12-14 \end{bmatrix}$$

$$\begin{bmatrix} -10 & -5 \\ -6 & 2 \\ 8 & -26 \end{bmatrix}$$

c. This example introduces parentheses, which means to perform addition before scalar multiplication.

$\frac{1}{2}(C + D)$

$$\frac{1}{2}\begin{bmatrix} -2+8 & 3+(-2) \\ 4+(-3) & -2+2 \\ 0+(-4) & 4+7 \end{bmatrix} = \frac{1}{2}\begin{bmatrix} 6 & 1 \\ 1 & 0 \\ -4 & 11 \end{bmatrix} = \begin{bmatrix} 3 & \frac{1}{2} \\ \frac{1}{2} & 0 \\ -2 & \frac{11}{2} \end{bmatrix}$$

What about all the other number properties? Do they apply to matrices? Can you distribute a scalar to matrices before performing the parentheses? In other words, is $\frac{1}{2}(C+D) = \frac{1}{2}C + \frac{1}{2}D$ a true statement? Will the distributive property work for any scalar, $k(A+B) = kA + kB$?

Let $A = [a_{ij}]$ and $B = [b_{ij}]$.

For $A \pm B = [a_{ij} \pm b_{ij}]$, $k(A \pm B) = [k(a_{ij} \pm b_{ij})] = [ka_{ij} \pm kb_{ij}] = kA \pm kB$.

Yes! Distribute property with scalar multiplication is true for matrices with the same order.

Matrix Multiplication

Given matrix $A = [a_{ij}]$ has order $m \times n$ and matrix $B = [b_{ij}]$ has order $n \times p$. The product $AB = [c_{ij}]$ is an $m \times p$ matrix, in which the n entries across the rows of A are multiplied with the n entries down the columns of B. These products are added to form the entries c_{ij}.

$$c_{ij} = a_{i1}b_{1j} + a_{i2}b_{2j} + a_{i3}b_{3j} + \ldots + a_{in}b_{nj}.$$

The product AB is defined only if the number of columns in A is equal to the number of rows in B.

Example 6: Multiply the following matrices, if possible.

a. $\begin{bmatrix} 2 & 3 & -2 \\ 3 & 1 & 6 \end{bmatrix} \begin{bmatrix} -1 \\ 2 \\ 4 \end{bmatrix}$ b. $\begin{bmatrix} -2 & 3 \\ 1 & 4 \end{bmatrix} \begin{bmatrix} 5 & 6 & -3 \\ 4 & 1 & 2 \end{bmatrix}$ c. $\begin{bmatrix} 6 & -3 & 9 \\ 12 & 0 & -6 \end{bmatrix} \begin{bmatrix} 6 & -3 & 9 \\ 12 & 0 & -6 \end{bmatrix}$

Solution:

a. First, determine the order of each matrix to determine if multiplication is possible.

$\begin{bmatrix} 2 & 3 & -2 \\ 3 & 1 & 6 \end{bmatrix} \begin{bmatrix} -1 \\ 2 \\ 4 \end{bmatrix}$ The number of columns in the 1st matrix equals the number of rows in the 2nd matrix.

2×3 3×1

The result will be a 2×1 matrix. Multiply the entries in row 1 by the entries in column 1, then add the products. The result is the 1st entry in the product.

$$\begin{bmatrix} 2 & 3 & -2 \\ 3 & 1 & 6 \end{bmatrix} \begin{bmatrix} -1 \\ 2 \\ 4 \end{bmatrix} = \begin{bmatrix} 2(-1)+3(2)+(-2)(4) \\ \end{bmatrix} = \begin{bmatrix} -4 \\ \end{bmatrix}$$

$$a_{11}b_{11} + a_{12}b_{21} + a_{13}b_{31}$$

Now, multiply the entries in row 2 by the entries in column 1. Then add the products. The result is the 2nd entry in the product.

$$\begin{bmatrix} 2 & 3 & -2 \\ 3 & 1 & 6 \end{bmatrix} \begin{bmatrix} -1 \\ 2 \\ 4 \end{bmatrix} = \begin{bmatrix} -4 \\ 3(-1)+1(2)+6(4) \end{bmatrix} = \begin{bmatrix} -4 \\ 23 \end{bmatrix}$$

$$a_{21}b_{11} + a_{22}b_{21} + a_{23}b_{31}$$

---

b. First, find the order of each matrix to determine if multiplication is possible.

$$\begin{bmatrix} -2 & 3 \\ 1 & 4 \end{bmatrix} \begin{bmatrix} 5 & 6 & -3 \\ 4 & 1 & 2 \end{bmatrix}$$ The number of columns in the 1st matrix equals the number of rows in the 2nd matrix.

$$2 \times 2 \qquad 2 \times 3$$

The result will be a 2×3 matrix. Multiply the entries in row 1 by the entries in each column, then add the products. The results are the entries in the first row of the product.

$$\begin{bmatrix} -2 & 3 \\ 1 & 4 \end{bmatrix} \begin{bmatrix} 5 & 6 & -3 \\ 4 & 1 & 2 \end{bmatrix} = \begin{bmatrix} -2(5)+3(4) & -2(6)+3(1) & -2(-3)+3(2) \\ & & \end{bmatrix} = \begin{bmatrix} 2 & -9 & 12 \\ & & \end{bmatrix}$$

$$a_{11}b_{11}+a_{12}b_{21} \qquad a_{11}b_{12}+a_{12}b_{22} \qquad a_{11}b_{13}+a_{12}b_{23}$$

Now, multiply the entries in row 2 by the entries in each column. Add the products. The results are the entries in the second row of the product.

$$\begin{bmatrix} -2 & 3 \\ 1 & 4 \end{bmatrix} \begin{bmatrix} 5 & 6 & -3 \\ 4 & 1 & 2 \end{bmatrix} = \begin{bmatrix} 2 & -9 & 12 \\ 1(5)+4(4) & 1(6)+4(1) & 1(-3)+4(2) \end{bmatrix} = \begin{bmatrix} 2 & -9 & 12 \\ 21 & 10 & 5 \end{bmatrix}$$

$$a_{21}b_{11}+a_{22}b_{21} \qquad a_{21}b_{12}+a_{22}b_{22} \qquad a_{21}b_{13}+a_{22}b_{23}$$

c. First, determine the order of each matrix to determine if multiplication is possible.

$$\begin{bmatrix} 6 & -3 & 9 \\ 12 & 0 & -6 \end{bmatrix} \begin{bmatrix} 6 & -3 & 9 \\ 12 & 0 & -6 \end{bmatrix}$$
$$\quad 2\times 3 \qquad\qquad 2\times 3$$

The number of columns in the 1st matrix does not equal the number of rows in the 2nd matrix. Therefore, matrix multiplication is not possible.

We found that matrix addition is commutative, just like regular number addition. What about matrix multiplication? Do you think matrix multiplication is commutative? Look at Example 6a. If we switch the order of the two matrices, will we get the same result?

Example 6a:

$$\begin{bmatrix} -1 \\ 2 \\ 4 \end{bmatrix} \begin{bmatrix} 2 & 3 & -2 \\ 3 & 1 & 6 \end{bmatrix}$$
$$3\times 1 \quad 2\times 3$$

The number of columns in the 1st matrix does not equal the number of rows in the 2nd matrix. Matrix multiplication is not possible, and the result is not the same as that found in the solution to part a. This means the commutative property does not exist for matrix multiplication. In other words, the order in which you multiply matrices is important!

Properties of Matrices

For matrices $A, B,$ and C, assuming that the indicated operations are possible:

1. Additive Identity $\qquad\qquad A+0=0+A=A$

2. Additive Inverse $\qquad\qquad A+(-A)=(-A)+A=0$

3. Commutative Property of Addition $\qquad A+B=B+A$

4. Associative Property of Addition $\qquad A+(B+C)=(A+B)+C$

5. Associative Property of Multiplication $\quad A(BC)=(AB)C$

6. Distributive Property $\qquad k(A+B)=kA+kB \qquad k$ is any scalar

$\qquad\qquad\qquad\qquad\qquad\quad A(B+C)=AB+AC$

$\qquad\qquad\qquad\qquad\qquad\quad (B+C)A=BA+CA$

One application to matrix operations is solving matrix equations. The actual solving process will be discussed in Lesson 12, but we can learn one important step in this lesson. Take a look at the following system of equations.

$$\begin{cases} 2x+3y-z=5 \\ x-2y+2z=6 \\ 4x+y-3z=5 \end{cases}$$

This system can be written as a matrix equation using matrix multiplication and the equality of matrices.

In general, a system of equations can be written as a matrix equation using the form $AX = B$, where matrix A is the **coefficient matrix**, X is the variable matrix, and B is the constants.

For this system, the equation $AX = B$ would be

$$\begin{bmatrix} 2 & 3 & -1 \\ 1 & -2 & 2 \\ 4 & 1 & -3 \end{bmatrix} \begin{bmatrix} x \\ y \\ z \end{bmatrix} = \begin{bmatrix} 5 \\ 6 \\ 5 \end{bmatrix}$$
$\quad\; 3\times 3 \qquad\quad 3\times 1 \quad\; 3\times 1$

Notice, matrix multiplication is possible with A and X because the number of columns in A equals the number of rows in X. The result of the multiplication would be 3×1, which is the order of matrix B.

$$\begin{bmatrix} 2 & 3 & -1 \\ 1 & -2 & 2 \\ 4 & 1 & -3 \end{bmatrix} \begin{bmatrix} x \\ y \\ z \end{bmatrix} = \begin{bmatrix} 5 \\ 6 \\ 5 \end{bmatrix}$$

If we were to multiply A and X, the result is $\begin{bmatrix} 2x+3y-z \\ x-2y+2z \\ 4x+y-3z \end{bmatrix} = \begin{bmatrix} 5 \\ 6 \\ 5 \end{bmatrix}$

$\qquad\qquad\qquad\qquad\qquad\qquad\qquad\qquad\quad\; 3\times 1 \qquad 3\times 1$

Now use the equality of matrices. The result is the original equations of the system.

$2x+3y-z=5$

$x-2y+2z=6$

$4x+y-3z=5$

Lesson 11 Practice Exercises

In Exercises 1 – 3, determine the missing entries to create a true statement.

1. $\begin{bmatrix} 2 & k \\ -3 & 0 \end{bmatrix} = \begin{bmatrix} a & -8 \\ -3 & 0 \end{bmatrix}$

2. $\begin{bmatrix} 9 \\ 2x-1 \\ -9 \\ 3 \end{bmatrix} = \begin{bmatrix} 9 \\ 0 \\ -9 \\ -y \end{bmatrix}$

3. $\begin{bmatrix} 5x & 1 \\ -7 & 4 \\ -3y & 6 \end{bmatrix} = \begin{bmatrix} 5 & 1 \\ -7 & b \\ 8 & r+2 \end{bmatrix}$

In Exercises 4 and 5, find the additive inverse matrix of the given matrix.

4. $A = \begin{bmatrix} 7 & -\frac{1}{2} & 9 \\ 0 & 1 & -6 \\ -1 & 2 & \frac{2}{5} \end{bmatrix}$

5. $B = \begin{bmatrix} -5 & 3 & 1.4 \\ 6.3 & -1.1 & 0 \end{bmatrix}$

In Exercises 6 – 20, perform the indicated operation.

$A = \begin{bmatrix} -4 & 9 \\ 11 & -3 \end{bmatrix}$

$B = \begin{bmatrix} 7 & 20 \\ -10 & 0 \end{bmatrix}$

$C = \begin{bmatrix} 2 \\ -1 \\ 1 \end{bmatrix}$

$D = \begin{bmatrix} 0 & -9 & 3 \\ 1 & 4 & -7 \end{bmatrix}$

$E = \begin{bmatrix} 13 & 1 & -1 \\ 9 & -14 & -2 \end{bmatrix}$

6. $A + B$

7. $D - E$

8. AB

9. BD

10. $C + E$

11. $4D$

12. DB

13. $A + D$

14. $-5C$

15. A^2

16. $3A - 2B$

17. $4E + 2D$

18. $3(B - A)$

19. EC

20. BA

In Exercises 21 – 24, determine if the following statements are true or false for matrices. Assume each matrix has order 2×2.

21. $A + B = B + A$

22. $A - B = B - A$

23. $AB = BA$

24. $A(B + C) = AB + AC$

In Exercises 25 – 27, write a matrix equation equivalent to the system of equations. State the order of each matrix in the equation.

25. $\begin{cases} 3x+2y-z=-4 \\ 2x-3y-2z=2 \\ x-y+2z=9 \end{cases}$

26. $\begin{cases} -5x+y=-13 \\ 6x+2y=22 \end{cases}$

27. $\begin{cases} w+2x+3y-4z=-10 \\ -3w+x-2y=6 \\ 2w+3x+5z=1 \\ -5w-x+y+2z=21 \end{cases}$

Solutions to Practice Exercises

Lesson 11

1. $a=2; k=-8$

2. $x=\dfrac{1}{2}; y=-3$

3. $x=1; y=-\dfrac{8}{3}; b=4; r=4$

4. $-A = \begin{bmatrix} -7 & \frac{1}{2} & -9 \\ 0 & -1 & 6 \\ 1 & -2 & -\frac{2}{5} \end{bmatrix}$

5. $-B = \begin{bmatrix} 5 & -3 & -1.4 \\ -6.3 & 1.1 & 0 \end{bmatrix}$

6. $\begin{bmatrix} 3 & 29 \\ 1 & -3 \end{bmatrix}$

7. $\begin{bmatrix} -13 & -10 & 4 \\ -8 & 18 & -5 \end{bmatrix}$

8. $\begin{bmatrix} -118 & -80 \\ 107 & 220 \end{bmatrix}$

9. $\begin{bmatrix} 20 & 17 & -119 \\ 0 & 90 & -30 \end{bmatrix}$

10. Not Possible

11. $\begin{bmatrix} 0 & -36 & 12 \\ 4 & 16 & -28 \end{bmatrix}$

12. Not Possible

13. Not Possible

14. $\begin{bmatrix} -10 \\ 5 \\ -5 \end{bmatrix}$

15. $\begin{bmatrix} 115 & -63 \\ -77 & 108 \end{bmatrix}$

16. $\begin{bmatrix} -26 & -13 \\ 53 & -9 \end{bmatrix}$

17. $\begin{bmatrix} 52 & -14 & 2 \\ 38 & -48 & -22 \end{bmatrix}$

18. $\begin{bmatrix} 33 & 33 \\ -63 & 9 \end{bmatrix}$

19. $\begin{bmatrix} 24 \\ 30 \end{bmatrix}$

20. $\begin{bmatrix} 192 & 3 \\ 40 & -90 \end{bmatrix}$

21. True

22. False

23. False

24. True

25. $\begin{bmatrix} 3 & 2 & -1 \\ 2 & -3 & -2 \\ 1 & -1 & 2 \end{bmatrix} \begin{bmatrix} x \\ y \\ z \end{bmatrix} = \begin{bmatrix} -4 \\ 2 \\ 9 \end{bmatrix}$

26. $\begin{bmatrix} -5 & 1 \\ 6 & 2 \end{bmatrix} \begin{bmatrix} x \\ y \end{bmatrix} = \begin{bmatrix} -13 \\ 22 \end{bmatrix}$

27. $\begin{bmatrix} 1 & 2 & 3 & -4 \\ -3 & 1 & -2 & 0 \\ 2 & 3 & 0 & 5 \\ -5 & -1 & 1 & 2 \end{bmatrix} \begin{bmatrix} w \\ x \\ y \\ z \end{bmatrix} = \begin{bmatrix} -10 \\ 6 \\ 1 \\ 21 \end{bmatrix}$

Lesson 12 — Matrix Inverse

A square matrix that has order $n \times n$ whose entries on the main diagonal are 1 and 0 elsewhere is called the **identity** matrix, I_n.

For a 2×2, the identity matrix is $I_2 = \begin{bmatrix} 1 & 0 \\ 0 & 1 \end{bmatrix}$.

For a 3×3, the identity matrix is $I_3 = \begin{bmatrix} 1 & 0 & 0 \\ 0 & 1 & 0 \\ 0 & 0 & 1 \end{bmatrix}$.

A special property of the identity matrix includes multiplication. If matrix multiplication is possible, then $IA = A$ and $AI = A$.

Example 1: Verify $IA = A$ and $AI = A$ is true for $A = \begin{bmatrix} 2 & 7 \\ 1 & 4 \end{bmatrix}$.

Solution:

$$\begin{bmatrix} 1 & 0 \\ 0 & 1 \end{bmatrix} \begin{bmatrix} 2 & 7 \\ 1 & 4 \end{bmatrix} = \begin{bmatrix} 1(2)+0(1) & 1(7)+0(4) \\ 0(2)+1(1) & 0(7)+1(4) \end{bmatrix} = \begin{bmatrix} 2 & 7 \\ 1 & 4 \end{bmatrix}$$

and

$$\begin{bmatrix} 2 & 7 \\ 1 & 4 \end{bmatrix} \begin{bmatrix} 1 & 0 \\ 0 & 1 \end{bmatrix} = \begin{bmatrix} 2(1)+7(0) & 2(0)+7(1) \\ 1(1)+4(0) & 1(0)+4(1) \end{bmatrix} = \begin{bmatrix} 2 & 7 \\ 1 & 4 \end{bmatrix}$$

Let A and B be square matrices with order $n \times n$. If $AB = I_n$ and $BA = I_n$, then matrix B is called the **inverse** of matrix A. Matrix B is renamed A^{-1}.

Example 2: Verify that matrix B is the inverse of matrix A using the definition.
$A = \begin{bmatrix} 2 & 7 \\ 1 & 4 \end{bmatrix}$ and $B = \begin{bmatrix} 4 & -7 \\ -1 & 2 \end{bmatrix}$

Solution: Determine if matrix multiplication results in the identity matrix.

Show $AB = I_n$.

$$\begin{bmatrix} 2 & 7 \\ 1 & 4 \end{bmatrix} \begin{bmatrix} 4 & -7 \\ -1 & 2 \end{bmatrix} = \begin{bmatrix} 2(4)+7(-1) & 2(-7)+7(2) \\ 1(4)+4(-1) & 1(-7)+4(2) \end{bmatrix} = \begin{bmatrix} 1 & 0 \\ 0 & 1 \end{bmatrix}$$

Now, show $BA = I_n$.

$$\begin{bmatrix} 4 & -7 \\ -1 & 2 \end{bmatrix} \begin{bmatrix} 2 & 7 \\ 1 & 4 \end{bmatrix} = \begin{bmatrix} 4(2)+(-7)(1) & 4(7)+(-7)(4) \\ (-1)(2)+2(1) & (-1)(7)+2(4) \end{bmatrix} = \begin{bmatrix} 1 & 0 \\ 0 & 1 \end{bmatrix}$$

Therefore matrix B is renamed $A^{-1} = \begin{bmatrix} 4 & -7 \\ -1 & 2 \end{bmatrix}$.

Steps for finding an inverse matrix:

1. Form an augmented matrix with the identity on the right side. $[A | I]$

2. Perform row operations to get the matrix in reduced row-echelon form (RREF). In other words, use Gauss-Jordan elimination. This puts the identity on the left side and the inverse on the right.

$$[I | A^{-1}]$$

Example 3: Find the inverse of the following matrices.

a. $A = \begin{bmatrix} -2 & -6 \\ 4 & 11 \end{bmatrix}$ **b.** $A = \begin{bmatrix} 1 & -1 & 2 \\ 2 & 0 & 6 \\ 3 & -5 & 7 \end{bmatrix}$

Solution:

Start with $[A | I]$.

a. $\begin{bmatrix} -2 & -6 & | & 1 & 0 \\ 4 & 11 & | & 0 & 1 \end{bmatrix}$ $-\dfrac{1}{2} R1 \to R1$

$\begin{bmatrix} 1 & 3 & | & -\dfrac{1}{2} & 0 \\ 4 & 11 & | & 0 & 1 \end{bmatrix}$ $-4R1 + R2 \to R2$

$\begin{bmatrix} 1 & 3 & | & -\dfrac{1}{2} & 0 \\ 0 & -1 & | & 2 & 1 \end{bmatrix}$ $-R2 \to R2$

Go to the MATRIX menu. For a TI-84, press 2ND and x^{-1}. Move the cursor to EDIT and press enter. For Example 3a, change the order of the matrix to 2×2 and enter the numbers in the matrix.

Return to the home screen by pressing 2ND and MODE. Now you are ready to use the matrix to perform any matrix operation. Press 2ND and x^{-1}, but this time move the cursor to NAMES and press 1 for matrix A. To find the inverse of A, press x^{-1}.

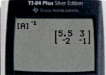

$$\begin{bmatrix} 1 & 3 & | & -\frac{1}{2} & 0 \\ 0 & 1 & | & -2 & -1 \end{bmatrix} \quad -3R2 + R1 \to R1$$

$$\begin{bmatrix} 1 & 0 & | & \frac{11}{2} & 3 \\ 0 & 1 & | & -2 & -1 \end{bmatrix} \quad A^{-1} = \begin{bmatrix} \frac{11}{2} & 3 \\ -2 & -1 \end{bmatrix}$$

---

b. $\begin{bmatrix} 1 & -1 & 2 & | & 1 & 0 & 0 \\ 2 & 0 & 6 & | & 0 & 1 & 0 \\ 3 & -5 & 7 & | & 0 & 0 & 1 \end{bmatrix}$ $\quad -2R1 + R2 \to R2$
$\quad -3R1 + R3 \to R3$

Since this matrix already has a 1 in the main diagonal position, we need 0's below. Multiply row 1 by –2 and add row 2. Multiply row 1 by –3 and add row 3.

$$\begin{bmatrix} 1 & -1 & 2 & | & 1 & 0 & 0 \\ 0 & 2 & 2 & | & -2 & 1 & 0 \\ 0 & -2 & 1 & | & -3 & 0 & 1 \end{bmatrix} \quad \frac{1}{2}R2 \to R2$$

To get a matrix in RREF, we move to column 2. Get a 1 in the main diagonal position by multiplying by 1/2.

$$\begin{bmatrix} 1 & -1 & 2 & | & 1 & 0 & 0 \\ 0 & 1 & 1 & | & -1 & \frac{1}{2} & 0 \\ 0 & -2 & 1 & | & -3 & 0 & 1 \end{bmatrix} \quad \begin{array}{l} R2 + R1 \to R1 \\ \\ 2R2 + R3 \to R3 \end{array}$$

Now get 0's above and below the main diagonal in column 2. Add row 2 and row 1. Multiply row 2 by 2 and add row 3.

$$\begin{bmatrix} 1 & 0 & 3 & | & 0 & \frac{1}{2} & 0 \\ 0 & 1 & 1 & | & -1 & \frac{1}{2} & 0 \\ 0 & 0 & 3 & | & -5 & 1 & 1 \end{bmatrix} \quad \frac{1}{3}R3 \to R3$$

Column 1 and 2 are complete. Move to column 3. Get a 1 in the main diagonal position by multiplying by 1/3.

$$\begin{bmatrix} 1 & 0 & 3 & | & 0 & \frac{1}{2} & 0 \\ 0 & 1 & 1 & | & -1 & \frac{1}{2} & 0 \\ 0 & 0 & 1 & | & -\frac{5}{3} & \frac{1}{3} & \frac{1}{3} \end{bmatrix} \quad \begin{array}{l} -3R3 + R1 \to R1 \\ \\ -1R3 + R2 \to R2 \end{array}$$

Now get 0's above the main diagonal in column 3. Multiply row 3 by –3 and add row 1. Multiply row 3 by –1 and add row 2.

$$\begin{bmatrix} 1 & 0 & 0 & | & 5 & -\frac{1}{2} & -1 \\ 0 & 1 & 0 & | & \frac{2}{3} & \frac{1}{6} & -\frac{1}{3} \\ 0 & 0 & 1 & | & -\frac{5}{3} & \frac{1}{3} & \frac{1}{3} \end{bmatrix} \quad A^{-1} = \begin{bmatrix} 5 & -\frac{1}{2} & -1 \\ \frac{2}{3} & \frac{1}{6} & -\frac{1}{3} \\ -\frac{5}{3} & \frac{1}{3} & \frac{1}{3} \end{bmatrix}$$

Check your answer by multiplying $AA^{-1} = I$ and $A^{-1}A = I$. The calculator is also a useful tool to check to make sure all your steps are correct.

Not every matrix has an inverse. First, the matrix must be square. Therefore, a matrix with order $m \times n$ will not have an inverse. Also, some square matrices do not have inverses. This can be determined when the final row of Gauss-Jordan elimination results in 0's. If a matrix does not have an inverse, the matrix is called **singular**. If a matrix has an inverse, the matrix is called **nonsingular** (or invertible matrix).

Example 4: Determine if the following matrix is nonsingular or singular.

$$A = \begin{bmatrix} -3 & 1 & -1 \\ 1 & -4 & -7 \\ 1 & 2 & 5 \end{bmatrix}$$

Solution: Start with $[A|I]$. Perform row operations to get the identity matrix on the left side.

$$\begin{bmatrix} -3 & 1 & -1 & | & 1 & 0 & 0 \\ 1 & -4 & -7 & | & 0 & 1 & 0 \\ 1 & 2 & 5 & | & 0 & 0 & 1 \end{bmatrix} \quad \text{Rowswap } R1 \leftrightarrow R1$$

$$\begin{bmatrix} 1 & -4 & -7 & | & 0 & 1 & 0 \\ -3 & 1 & -1 & | & 1 & 0 & 0 \\ 1 & 2 & 5 & | & 0 & 0 & 1 \end{bmatrix} \quad \begin{array}{l} 3R1 + R2 \rightarrow R2 \\ -R1 + R3 \rightarrow R3 \end{array}$$

$$\begin{bmatrix} 1 & -4 & -7 & | & 0 & 1 & 0 \\ 0 & -11 & -22 & | & 1 & 3 & 0 \\ 0 & 6 & 12 & | & 0 & -1 & 1 \end{bmatrix} \quad -\frac{1}{11} R2 \rightarrow R2$$

$$\begin{bmatrix} 1 & -4 & -7 & | & 0 & 1 & 0 \\ 0 & 1 & 2 & | & -\frac{1}{11} & -\frac{3}{11} & 0 \\ 0 & 6 & 12 & | & 0 & -1 & 1 \end{bmatrix} \quad \begin{array}{l} 4R2 + R1 \rightarrow R1 \\ -6R2 + R3 \rightarrow R3 \end{array}$$

$$\begin{bmatrix} 1 & 0 & 1 & | & -\frac{4}{11} & -\frac{1}{11} & 0 \\ 0 & 1 & 2 & | & -\frac{1}{11} & -\frac{3}{11} & 0 \\ 0 & 0 & 0 & | & \frac{6}{11} & \frac{7}{11} & 1 \end{bmatrix}$$

It is impossible to get a 1 in the main diagonal entry a_{33}.

Since the left side has a final row of 0's, it is impossible to form the identity matrix. Therefore, this matrix is singular. Matrix A does not have an inverse.

At the end of Lesson 11, a system of equations was written as a matrix equation $AX = B$.

$$\begin{cases} a_1x + b_1y + c_1z = d_1 \\ a_2x + b_2y + c_2z = d_2 \\ a_3x + b_3y + c_3z = d_3 \end{cases}$$

Matrix A is the **coefficient matrix**. $A = \begin{bmatrix} a_1 & b_1 & c_1 \\ a_2 & b_2 & c_2 \\ a_3 & b_3 & c_3 \end{bmatrix}$

Matrix X is the variable matrix. Be sure it is written as a **column matrix**. $X = \begin{bmatrix} x \\ y \\ z \end{bmatrix}$

A column matrix has order $m \times 1$.

Matrix B is the constant matrix. Be sure it is also written as a column matrix. $B = \begin{bmatrix} d_1 \\ d_2 \\ d_3 \end{bmatrix}$

Example 5: Write the following system as a matrix equation.

$$\begin{cases} 3x + 4y = -1 \\ 3x + 5y = 1 \end{cases}$$

Solution: $AX = B$

$$\begin{bmatrix} 3 & 4 \\ 3 & 5 \end{bmatrix} \begin{bmatrix} x \\ y \end{bmatrix} = \begin{bmatrix} -1 \\ 1 \end{bmatrix}$$

Notice that normal algebra would tell us to divide both sides by A to solve for the variable X. Since there is no such thing as matrix division, this method would not work. We can use an inverse matrix to solve the equation. Instead of dividing by matrix A, multiply both sides of the equation by the inverse of A. The order of multiplication is important, so make sure A^{-1} is on the left of each matrix.

$\underline{A^{-1}}AX = \underline{A^{-1}}B$ By definition of inverse, $A^{-1}A = I$.

$IX = A^{-1}B$ According to the identity matrix, any multiplication would result in the same matrix. $IX = X$

$X = A^{-1}B$ To find the value of each variable, multiply matrices A^{-1} and B.

Example 6: Solve the following systems using a matrix equation.

a. $\begin{cases} 3x+4y=-1 \\ 3x+5y=1 \end{cases}$

b. $\begin{cases} x+7z=20 \\ 2x+y-z=-3 \\ 7x+3y+z=2 \end{cases}$

Solution: Set up a matrix equation $AX = B$.

a. $\begin{bmatrix} 3 & 4 \\ 3 & 5 \end{bmatrix}\begin{bmatrix} x \\ y \end{bmatrix} = \begin{bmatrix} -1 \\ 1 \end{bmatrix}$ First, find the inverse of matrix $A = \begin{bmatrix} 3 & 4 \\ 3 & 5 \end{bmatrix}$.

Augment matrix A with the identity matrix. $\begin{bmatrix} 3 & 4 & | & 1 & 0 \\ 3 & 5 & | & 0 & 1 \end{bmatrix}$

Gauss-Jordan Elimination results in $\begin{bmatrix} 1 & 0 & | & \frac{5}{3} & -\frac{4}{3} \\ 0 & 1 & | & -1 & 1 \end{bmatrix}$, therefore $A^{-1} = \begin{bmatrix} \frac{5}{3} & -\frac{4}{3} \\ -1 & 1 \end{bmatrix}$.

Second, multiply both sides of the equation by the inverse.

$\begin{bmatrix} \frac{5}{3} & -\frac{4}{3} \\ -1 & 1 \end{bmatrix}\begin{bmatrix} 3 & 4 \\ 3 & 5 \end{bmatrix}\begin{bmatrix} x \\ y \end{bmatrix} = \begin{bmatrix} \frac{5}{3} & -\frac{4}{3} \\ -1 & 1 \end{bmatrix}\begin{bmatrix} -1 \\ 1 \end{bmatrix}$

$\begin{bmatrix} 1 & 0 \\ 0 & 1 \end{bmatrix}\begin{bmatrix} x \\ y \end{bmatrix} = \begin{bmatrix} \frac{5}{3} & -\frac{4}{3} \\ -1 & 1 \end{bmatrix}\begin{bmatrix} -1 \\ 1 \end{bmatrix}$ Multiply each row by the column matrix.

$\begin{bmatrix} x \\ y \end{bmatrix} = \begin{bmatrix} \frac{5}{3}(-1) + \left(-\frac{4}{3}\right)(1) \\ (-1)(-1) + 1(1) \end{bmatrix}$

$\begin{bmatrix} x \\ y \end{bmatrix} = \begin{bmatrix} -3 \\ 2 \end{bmatrix}$ From Lesson 11, the two matrices are equivalent if they have the same order and corresponding entries are equal. Therefore, $x = -3$ and $y = 2$.

The solution is $(-3, 2)$.

---

b. $\begin{bmatrix} 1 & 0 & 7 \\ 2 & 1 & -1 \\ 7 & 3 & 1 \end{bmatrix}\begin{bmatrix} x \\ y \\ z \end{bmatrix} = \begin{bmatrix} 20 \\ -3 \\ 2 \end{bmatrix}$

First, find the inverse of matrix $A = \begin{bmatrix} 1 & 0 & 7 \\ 2 & 1 & -1 \\ 7 & 3 & 1 \end{bmatrix}$.

Augment matrix A with the identity matrix.
$$\begin{bmatrix} 1 & 0 & 7 & | & 1 & 0 & 0 \\ 2 & 1 & -1 & | & 0 & 1 & 0 \\ 7 & 3 & 1 & | & 0 & 0 & 1 \end{bmatrix}$$

Gauss-Jordan Elimination results in $\begin{bmatrix} 1 & 0 & 0 & | & -\frac{4}{3} & -7 & \frac{7}{3} \\ 0 & 1 & 0 & | & 3 & 16 & -5 \\ 0 & 0 & 1 & | & \frac{1}{3} & 1 & -\frac{1}{3} \end{bmatrix}$, therefore $A^{-1} = \begin{bmatrix} -\frac{4}{3} & -7 & \frac{7}{3} \\ 3 & 16 & -5 \\ \frac{1}{3} & 1 & -\frac{1}{3} \end{bmatrix}$.

Second, multiply both sides the equation by the inverse.

$$\begin{bmatrix} -\frac{4}{3} & -7 & \frac{7}{3} \\ 3 & 16 & -5 \\ \frac{1}{3} & 1 & -\frac{1}{3} \end{bmatrix} \begin{bmatrix} 1 & 0 & 7 \\ 2 & 1 & -1 \\ 7 & 3 & 1 \end{bmatrix} \begin{bmatrix} x \\ y \\ z \end{bmatrix} = \begin{bmatrix} -\frac{4}{3} & -7 & \frac{7}{3} \\ 3 & 16 & -5 \\ \frac{1}{3} & 1 & -\frac{1}{3} \end{bmatrix} \begin{bmatrix} 20 \\ -3 \\ 2 \end{bmatrix}$$

$A^{-1}A = I \longrightarrow \begin{bmatrix} 1 & 0 & 0 \\ 0 & 1 & 0 \\ 0 & 0 & 1 \end{bmatrix} \begin{bmatrix} x \\ y \\ z \end{bmatrix} = \begin{bmatrix} -\frac{4}{3} & -7 & \frac{7}{3} \\ 3 & 16 & -5 \\ \frac{1}{3} & 1 & -\frac{1}{3} \end{bmatrix} \begin{bmatrix} 20 \\ -3 \\ 2 \end{bmatrix}$ Multiply each row by the column matrix.

$$\begin{bmatrix} x \\ y \\ z \end{bmatrix} = \begin{bmatrix} -\frac{4}{3}(20) + (-7)(-3) + \frac{7}{3}(2) \\ 3(20) + 16(-3) + (-5)(2) \\ \frac{1}{3}(20) + 1(-3) + \left(-\frac{1}{3}\right)(2) \end{bmatrix}$$

$$\begin{bmatrix} x \\ y \\ z \end{bmatrix} = \begin{bmatrix} -1 \\ 2 \\ 3 \end{bmatrix}$$

The solution is $(-1, 2, 3)$.

Remember to check your solutions by evaluating the ordered pair or triple for each equation in the system. For Example 6b, substitute $x = -1$, $y = 2$, and $z = 3$ into all three equations.

$$\begin{cases} (-1) + 7(3) = 20 \quad \checkmark \\ 2(-1) + (2) - (3) = -3 \quad \checkmark \\ 7(-1) + 3(2) + (3) = 2 \quad \checkmark \end{cases}$$

Lesson 12　　　　　　　　　　　　　　　　　Practice Exercises

In Exercises 1 – 6, find the inverse of the given matrix, if it exists. Determine if the matrix is singular or nonsingular. Use the calculator to verify the results.

1. $A = \begin{bmatrix} 4 & -3 \\ -2 & 1 \end{bmatrix}$
2. $A = \begin{bmatrix} 5 & 6 \\ 2 & -3 \end{bmatrix}$
3. $A = \begin{bmatrix} 3 & -4 \\ 6 & -8 \end{bmatrix}$

4. $A = \begin{bmatrix} 2 & 1 & -2 \\ 1 & 1 & 1 \\ 0 & -2 & -1 \end{bmatrix}$
5. $A = \begin{bmatrix} -2 & 3 & -5 \\ 5 & -7 & 12 \\ 1 & -1 & 2 \end{bmatrix}$
6. $A = \begin{bmatrix} 1 & 1 & -2 \\ 4 & -1 & 3 \\ 3 & 2 & -1 \end{bmatrix}$

In Exercises 7 and 8, verify the matrices are inverses by showing $AB = I$ and $BA = I$.

7. $A = \begin{bmatrix} -2 & -6 \\ 4 & 11 \end{bmatrix}$ and $B = \begin{bmatrix} 5.5 & 3 \\ -2 & -1 \end{bmatrix}$

8. $A = \begin{bmatrix} 1 & 2 & -1 \\ 1 & 0 & 1 \\ 2 & -1 & 1 \end{bmatrix}$ and $B = \begin{bmatrix} \frac{1}{4} & -\frac{1}{4} & \frac{1}{2} \\ \frac{1}{4} & \frac{3}{4} & -\frac{1}{2} \\ -\frac{1}{4} & \frac{5}{4} & -\frac{1}{2} \end{bmatrix}$

In Exercises 9 – 17, write a matrix equation equivalent to the system of equations. Find the inverse of the coefficient matrix. Then solve the matrix equation.

9. $\begin{cases} 2x + 4y = 6 \\ x - 4y = -12 \end{cases}$
10. $\begin{cases} -3x + 4y = -4 \\ 2x - y = 6 \end{cases}$
11. $\begin{cases} 3x - 2y = 4 \\ 9x + 4y = -3 \end{cases}$

12. $\begin{cases} x + y - z = 7 \\ 2x + y + z = 5 \\ -x + y + 2z = -10 \end{cases}$
13. $\begin{cases} 3x - y + z = 6 \\ 2x + 2y - z = 5 \\ 2x - y + z = 5 \end{cases}$
14. $\begin{cases} 3x - 4y + z = -20 \\ x + 2y - z = 16 \\ 2x - 3y + 2z = 0 \end{cases}$

15. $\begin{cases} -2x + y + 4z = -8 \\ 2x + y - 3z = 11 \\ x - y = 1 \end{cases}$
16. $\begin{cases} x + y = 2 \\ 3x + 2z = 5 \\ 2x + 3y - 3z = 9 \end{cases}$
17. $\begin{cases} -w + 2x - y - z = -1 \\ -4w + 9x - 5y - 6z = 1 \\ x - y - z = 4 \\ 3w - 5x + 3y + 3z = 2 \end{cases}$

Solutions for Practice Exercises   **Lesson 12**

1. $\begin{bmatrix} -\frac{1}{2} & -\frac{3}{2} \\ -1 & -2 \end{bmatrix}$; Nonsingular   2. $\begin{bmatrix} \frac{1}{9} & \frac{2}{9} \\ \frac{2}{27} & -\frac{5}{27} \end{bmatrix}$; Nonsingular   3. Singular Matrix

4. $\begin{bmatrix} \frac{1}{7} & \frac{5}{7} & \frac{3}{7} \\ \frac{1}{7} & -\frac{2}{7} & -\frac{4}{7} \\ -\frac{2}{7} & \frac{4}{7} & \frac{1}{7} \end{bmatrix}$; Nonsingular   5. Singular Matrix   6. $\begin{bmatrix} \frac{5}{14} & \frac{3}{14} & -\frac{1}{14} \\ -\frac{13}{14} & -\frac{5}{14} & \frac{11}{14} \\ -\frac{11}{14} & -\frac{1}{14} & \frac{5}{14} \end{bmatrix}$; Nonsingular

7. Verified   8. Verified

9. $\begin{bmatrix} 2 & 4 \\ 1 & -4 \end{bmatrix}\begin{bmatrix} x \\ y \end{bmatrix} = \begin{bmatrix} 6 \\ -12 \end{bmatrix}$ 10. $\begin{bmatrix} -3 & 4 \\ 2 & -1 \end{bmatrix}\begin{bmatrix} x \\ y \end{bmatrix} = \begin{bmatrix} -4 \\ 6 \end{bmatrix}$ 11. $\begin{bmatrix} 3 & -2 \\ 9 & 4 \end{bmatrix}\begin{bmatrix} x \\ y \end{bmatrix} = \begin{bmatrix} 4 \\ -3 \end{bmatrix}$

$A^{-1} = \begin{bmatrix} \frac{1}{3} & \frac{1}{3} \\ \frac{1}{12} & -\frac{1}{6} \end{bmatrix}$ $\left(-2, \frac{5}{2}\right)$ $A^{-1} = \begin{bmatrix} \frac{1}{5} & \frac{4}{5} \\ \frac{2}{5} & \frac{3}{5} \end{bmatrix}$ $(4,2)$ $A^{-1} = \begin{bmatrix} \frac{2}{15} & \frac{1}{15} \\ -\frac{3}{10} & \frac{1}{10} \end{bmatrix}$ $\left(\frac{1}{3}, -\frac{3}{2}\right)$

12. $\begin{bmatrix} 1 & 1 & -1 \\ 2 & 1 & 1 \\ -1 & 1 & 2 \end{bmatrix}\begin{bmatrix} x \\ y \\ z \end{bmatrix} = \begin{bmatrix} 7 \\ 5 \\ -10 \end{bmatrix}$ 13. $\begin{bmatrix} 3 & -1 & 1 \\ 2 & 2 & -1 \\ 2 & -1 & 1 \end{bmatrix}\begin{bmatrix} x \\ y \\ z \end{bmatrix} = \begin{bmatrix} 6 \\ 5 \\ 5 \end{bmatrix}$ 14. $\begin{bmatrix} 3 & -4 & 1 \\ 1 & 2 & -1 \\ 2 & -3 & 2 \end{bmatrix}\begin{bmatrix} x \\ y \\ z \end{bmatrix} = \begin{bmatrix} -20 \\ 16 \\ 0 \end{bmatrix}$

$A^{-1} = \begin{bmatrix} -\frac{1}{7} & \frac{3}{7} & -\frac{2}{7} \\ \frac{5}{7} & -\frac{1}{7} & \frac{3}{7} \\ -\frac{3}{7} & \frac{2}{7} & \frac{1}{7} \end{bmatrix}$ $(4,0,-3)$ $A^{-1} = \begin{bmatrix} 1 & 0 & -1 \\ -4 & 1 & 5 \\ -6 & 1 & 8 \end{bmatrix}$ $(1,6,9)$ $A^{-1} = \begin{bmatrix} \frac{1}{12} & \frac{5}{12} & \frac{1}{6} \\ -\frac{1}{3} & \frac{1}{3} & \frac{1}{3} \\ -\frac{7}{12} & \frac{1}{12} & \frac{5}{6} \end{bmatrix}$ $(5,12,13)$

15. $\begin{bmatrix} -2 & 1 & 4 \\ 2 & 1 & -3 \\ 1 & -1 & 0 \end{bmatrix}\begin{bmatrix} x \\ y \\ z \end{bmatrix} = \begin{bmatrix} -8 \\ 11 \\ 1 \end{bmatrix}$ 16. $\begin{bmatrix} 1 & 1 & 0 \\ 3 & 0 & 2 \\ 2 & 3 & -3 \end{bmatrix}\begin{bmatrix} x \\ y \\ z \end{bmatrix} = \begin{bmatrix} 2 \\ 5 \\ 9 \end{bmatrix}$ 17. $\begin{bmatrix} -1 & 2 & -1 & -1 \\ -4 & 9 & -5 & -6 \\ 0 & 1 & -1 & -1 \\ 3 & -5 & 3 & 3 \end{bmatrix}\begin{bmatrix} w \\ x \\ y \\ z \end{bmatrix} = \begin{bmatrix} -1 \\ 1 \\ 4 \\ 2 \end{bmatrix}$

$A^{-1} = \begin{bmatrix} \frac{1}{3} & \frac{4}{9} & \frac{7}{9} \\ \frac{1}{3} & \frac{4}{9} & -\frac{2}{9} \\ \frac{1}{3} & \frac{1}{9} & \frac{4}{9} \end{bmatrix}$ $A^{-1} = \begin{bmatrix} -\frac{6}{7} & \frac{3}{7} & \frac{2}{7} \\ \frac{13}{7} & -\frac{3}{7} & -\frac{2}{7} \\ \frac{9}{7} & -\frac{1}{7} & -\frac{3}{7} \end{bmatrix}$ $A^{-1} = \begin{bmatrix} 2 & 0 & 1 & 1 \\ 3 & 0 & 0 & 1 \\ -1 & 1 & -2 & 1 \\ 4 & -1 & 1 & 0 \end{bmatrix}$

$(3,2,-1)$ $(3,-1,-2)$ $(4,-1,-4,-1)$

Lesson 13 — Determinants

The **determinant** of a matrix is a value that can be computed from the entries of a square matrix. This value is used in many applications, some of which we will discuss in this section. The notation used for determinant is $\det(A)$.

For a 2×2 matrix $A = \begin{bmatrix} a & b \\ c & d \end{bmatrix}$, the determinant is found using the formula $\det(A) = ad - bc$.

Example 1a: Find the determinant of matrix $C = \begin{bmatrix} 9 & 1 \\ 8 & 2 \end{bmatrix}$.

Solution: Multiply each diagonal and then find the difference. The order of subtraction is important.
$$\det(C) = 9(2) - 1(8)$$
$$= 18 - 8$$
$$= 10$$

The determinant for a 2×2 can be used to find the inverse of the matrix. For matrix $A = \begin{bmatrix} a & b \\ c & d \end{bmatrix}$, $A^{-1} = \dfrac{1}{\det(A)} \begin{bmatrix} d & -b \\ -c & a \end{bmatrix}$. This formula only works for a 2×2 matrix. The main diagonal entries are switched, and the other diagonal values are changed to their opposites. Then multiply each entry by the scalar $\dfrac{1}{\det(A)}$.

Example 1b: Find the inverse of matrix $C = \begin{bmatrix} 9 & 1 \\ 8 & 2 \end{bmatrix}$.

Solution:

$$C^{-1} = \frac{1}{\det(C)} \begin{bmatrix} 2 & -1 \\ -8 & 9 \end{bmatrix} = \begin{bmatrix} \left(\frac{1}{10}\right)2 & \left(\frac{1}{10}\right)(-1) \\ \left(\frac{1}{10}\right)(-8) & \left(\frac{1}{10}\right)9 \end{bmatrix}$$

$$= \begin{bmatrix} \frac{1}{5} & -\frac{1}{10} \\ -\frac{4}{5} & \frac{9}{10} \end{bmatrix}$$

For a 3×3 matrix $A = \begin{bmatrix} a_{11} & a_{12} & a_{13} \\ a_{21} & a_{22} & a_{23} \\ a_{31} & a_{32} & a_{33} \end{bmatrix}$, the determinant is found using two methods.

1st method: Expand using minors.

A **minor** m_{ij} of an entry is found by using the determinant of the matrix formed by deleting the ith row and j^{th} column of that entry. For example, the minor for entry a_{11} is the determinant of the matrix found by deleting row 1 and column 1. . $m_{11} = a_{22}a_{33} - a_{32}a_{23}$

Example 2: Find the following minors for the matrix $A = \begin{bmatrix} 2 & -1 & 5 \\ 4 & 3 & -7 \\ 8 & -7 & 6 \end{bmatrix}$.

a. Minor for entry a_{11} **b.** Minor for entry a_{32}

Solution:

a. Delete row 1 and column 1, and find the determinant of the resulting matrix.

$A = \begin{bmatrix} 2 & -1 & 5 \\ 4 & 3 & -7 \\ 8 & -7 & 6 \end{bmatrix}$ $\begin{bmatrix} 3 & -7 \\ -7 & 6 \end{bmatrix}$ $m_{11} = 3(6) - (-7)(-7) = -31$

b. Delete row 3 and column 2, and find the determinant of the resulting matrix.

$A = \begin{bmatrix} 2 & -1 & 5 \\ 4 & 3 & -7 \\ 8 & -7 & 6 \end{bmatrix}$ $\begin{bmatrix} 2 & 5 \\ 4 & -7 \end{bmatrix}$ $m_{32} = 2(-7) - 4(5) = -34$

A **cofactor** of an entry is found by multiplying the minor by a power of -1. $(-1)^{i+j} m_{ij}$

The cofactor for entry a_{11} would be $(-1)^{1+1} m_{11} = (1)(-31) = -31$. The cofactor for entry a_{32} would be $(-1)^{3+2} m_{32} = (-1)(-34) = 34$. For entry a_{11}, the cofactor equals the minor. And for entry a_{32}, the cofactor is opposite the minor.

To find the determinant, select any row or column. To show an example, we will select row 1. Please note, you could select the other rows or any of the three columns and the same determinant would result. Multiply each entry in row 1 by its cofactor.

$$A = \begin{bmatrix} a_{11} & a_{12} & a_{13} \\ a_{21} & a_{22} & a_{23} \\ a_{31} & a_{32} & a_{33} \end{bmatrix} \qquad \det(A) = a_{11}(-1)^{1+1}m_{11} + a_{12}(-1)^{1+2}m_{12} + a_{13}(-1)^{1+3}m_{13}$$

cofactors

Example 3a: Find the determinant of matrix $A = \begin{bmatrix} 2 & -1 & 5 \\ 4 & 3 & -7 \\ 8 & -7 & 6 \end{bmatrix}$ by minor expansion.

Solution: Select any row or column. Expand using minors.

Expanding using Row 1:

$$A = \begin{bmatrix} 2 & -1 & 5 \\ 4 & 3 & -7 \\ 8 & -7 & 6 \end{bmatrix}$$

$$\det(A) = a_{11}(-1)^{1+1}m_{11} + a_{12}(-1)^{1+2}m_{12} + a_{13}(-1)^{1+3}m_{13}$$

$$= 2(-1)^{1+1}\begin{vmatrix} 3 & -7 \\ -7 & 6 \end{vmatrix} + (-1)(-1)^{1+2}\begin{vmatrix} 4 & -7 \\ 8 & 6 \end{vmatrix} + 5(-1)^{1+3}\begin{vmatrix} 4 & 3 \\ 8 & -7 \end{vmatrix}$$

$$= 2(1)(-31) + (-1)(-1)(80) + 5(1)(-52)$$

$$= (-62) + 80 + (-260)$$

$$= -242$$

Notice, the determinant is the same if we select column 3.

$$A = \begin{bmatrix} 2 & -1 & 5 \\ 4 & 3 & -7 \\ 8 & -7 & 6 \end{bmatrix}$$

$$\det(A) = +a_{13}m_{13} - a_{23}m_{23} + a_{33}m_{33}$$

$$= 5\begin{vmatrix} 4 & 3 \\ 8 & -7 \end{vmatrix} - (-7)\begin{vmatrix} 2 & -1 \\ 8 & -7 \end{vmatrix} + 6\begin{vmatrix} 2 & -1 \\ 4 & 3 \end{vmatrix}$$

$$= 5(-52) - (-7)(-6) + (6)(10)$$

$$= (-260) + (-42) + (60)$$

$$= -242$$

When finding the cofactor, each minor is multiplied by the entry $(-1)^{i+j}$. This causes the signs used between terms of the expansion to alternate. For entry a_{11}, we will use $(-1)^{1+1} = 1$. For entry a_{12}, we will use $(-1)^{1+2} = -1$. The sign chart can be used instead of $(-1)^{i+j}$.

$$A = \begin{bmatrix} + & - & + \\ - & + & - \\ + & - & + \end{bmatrix}$$

For column 3, we would use + − +.

2nd Method: Column Rotation

Write the first two columns at the end of the matrix. Be sure to keep them in the same order.

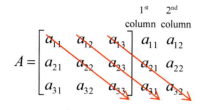

Multiply each main diagonal and add the products.

$a_{11}a_{22}a_{33} + a_{12}a_{23}a_{31} + a_{13}a_{21}a_{32}$

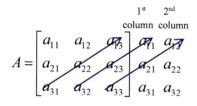

Multiply the other diagonals and add the products.

$a_{31}a_{22}a_{13} + a_{32}a_{23}a_{11} + a_{33}a_{21}a_{12}$

The final step is to find the difference. Be careful! The order matters!

$\det(A) = (a_{11}a_{22}a_{33} + a_{12}a_{23}a_{31} + a_{13}a_{21}a_{32}) - (a_{31}a_{22}a_{13} + a_{32}a_{23}a_{11} + a_{33}a_{21}a_{12})$

Example 3b: Find the determinant of matrix $A = \begin{bmatrix} 2 & -1 & 5 \\ 4 & 3 & -7 \\ 8 & -7 & 6 \end{bmatrix}$ using column rotation.

Solution: Start by writing the first two columns at the end of the matrix. Then multiply the entries in the diagonals.

$$\det(A) = [2(3)(6) + (-1)(-7)(8) + 5(4)(-7)] - [8(3)(5) + (-7)(-7)(2) + 6(4)(-1)]$$
$$= (-48) - 194$$
$$= -242$$

Expand using minors would work for higher order matrices. Column rotation only works for 3×3.

From Lesson 12, we learned that a nonsingular matrix has an inverse and a singular matrix does not have an inverse. If the determinant equals zero, the matrix will not have an inverse and is considered to be singular. If the determinant does not equal zero, the matrix will have an inverse and is considered nonsingular.

> **Example 4:** Determine if the matrix is nonsingular or singular using determinants.
>
> a. $B = \begin{bmatrix} -3 & 2 \\ -6 & 4 \end{bmatrix}$
> b. $D = \begin{bmatrix} 3 & -4 & -1 \\ 2 & -1 & 1 \\ 4 & -7 & -3 \end{bmatrix}$
>
> **Solution:**
>
> a. $\det(B) = (-3)(4) - (-6)(2)$
> $= -12 + 12$
> $= 0$
> Matrix B is singular.
>
> b. Using expansion by minors or column rotation, the determinant of matrix D is $\det(D) = 0$. Therefore, matrix D is singular.

We learned an application for the determinant of a 2×2 matrix in Example 1b. Now, we will learn three more applications for determinants: Cramer's Rule, Area of a Triangle, and Collinear Points.

Cramer's Rule for 2×2

$a_1 x + b_1 y = c_1$
$a_2 x + b_2 y = c_2$

$x = \dfrac{D_x}{D} \qquad y = \dfrac{D_y}{D}$

D is the determinant of the coefficient matrix, $\begin{bmatrix} a_1 & b_1 \\ a_2 & b_2 \end{bmatrix}$.

If $D = 0$, the coefficient matrix is singular. Since you cannot divide by zero, Cramer's Rule does not work for singular matrices. You would need to perform Gaussian or Gauss-Jordan elimination to determine if the system is inconsistent or dependent.

D_x is the determinant of the matrix, $\begin{bmatrix} c_1 & b_1 \\ c_2 & b_2 \end{bmatrix}$. Notice, the constants replace the column for the x values.

D_y is the determinant of the matrix, $\begin{bmatrix} a_1 & c_1 \\ a_2 & c_2 \end{bmatrix}$. Notice, the constants replace the column for the y values.

Cramer's Rule for 3×3

$$a_1x+b_1y+c_1z = d_1$$
$$a_2x+b_2y+c_2z = d_2 \qquad x = \frac{D_x}{D} \qquad y = \frac{D_y}{D} \qquad z = \frac{D_z}{D}$$
$$a_3x+b_3y+c_3z = d_3$$

D is the determinant of the coefficient matrix, $\begin{bmatrix} a_1 & b_1 & c_1 \\ a_2 & b_2 & c_2 \\ a_3 & b_3 & c_3 \end{bmatrix}$.

D_x is the determinant of the matrix, $\begin{bmatrix} d_1 & b_1 & c_1 \\ d_2 & b_2 & c_2 \\ d_3 & b_3 & c_3 \end{bmatrix}$. Notice, the constants replace the column for the x values.

D_y is the determinant of the matrix, $\begin{bmatrix} a_1 & d_1 & c_1 \\ a_2 & d_2 & c_2 \\ a_3 & d_3 & c_3 \end{bmatrix}$. Notice, the constants replace the column for the y values.

D_z is the determinant of the matrix, $\begin{bmatrix} a_1 & b_1 & d_1 \\ a_2 & b_2 & d_2 \\ a_3 & b_3 & d_3 \end{bmatrix}$. Notice, the constants replace the column for the z values.

Example 5: Solve the following systems using Cramer's Rule.

a. $\begin{cases} 5x-3y = 6 \\ 2x+4y = -7 \end{cases}$

b. $\begin{cases} 2x-3y+z = 2 \\ 4x+2z = -3 \\ 3x+y-2z = 1 \end{cases}$

Solution:

a. The coefficient matrix is $\begin{bmatrix} 5 & -3 \\ 2 & 4 \end{bmatrix}$. The determinant of the coefficient matrix is

$$D = 5(4) - 2(-3)$$
$$= 26$$

The matrix formed by replacing the constants for the x-coefficients is $\begin{bmatrix} 6 & -3 \\ -7 & 4 \end{bmatrix}$. The determinant

141

is $D_x = 6(4) - (-7)(-3) = 3$. The matrix formed by replacing the constants for the y-coefficients

is $\begin{bmatrix} 5 & 6 \\ 2 & -7 \end{bmatrix}$. The determinant is $D_y = 5(-7) - 2(6) = -47$. The solution to the system is

$x = \dfrac{D_x}{D} = \dfrac{3}{26}$ and $y = \dfrac{D_y}{D} = \dfrac{-47}{26}$, and the final answer is written as an ordered pair $\left(\dfrac{3}{26}, -\dfrac{47}{26}\right)$.

---

b. The coefficient matrix is $\begin{bmatrix} 2 & -3 & 1 \\ 4 & 0 & 2 \\ 3 & 1 & -2 \end{bmatrix}$. The determinant of the coefficient matrix is $D = -42$.

The matrix formed by replacing the constants for the x-coefficients is $\begin{bmatrix} 2 & -3 & 1 \\ -3 & 0 & 2 \\ 1 & 1 & -2 \end{bmatrix}$. The

determinant is $D_x = 5$. The matrix formed by replacing the constants for the y-coefficients is

$\begin{bmatrix} 2 & 2 & 1 \\ 4 & -3 & 2 \\ 3 & 1 & -2 \end{bmatrix}$. The determinant is $D_y = 49$. The matrix formed by replacing the constants for

the z-coefficients is $\begin{bmatrix} 2 & -3 & 2 \\ 4 & 0 & -3 \\ 3 & 1 & 1 \end{bmatrix}$. The determinant is $D_z = 53$.

$x = \dfrac{D_x}{D} = \dfrac{5}{-42}$, $y = \dfrac{D_y}{D} = \dfrac{49}{-42}$, and $z = \dfrac{D_z}{D} = \dfrac{53}{-42}$.

The final answer written as an ordered triple is $\left(-\dfrac{5}{42}, -\dfrac{7}{6}, -\dfrac{53}{42}\right)$.

Area of a Triangle

For a triangle with vertices (x_1, y_1), (x_2, y_2), and (x_3, y_3), the area of the triangle is

$A = \dfrac{|\det(T)|}{2}$, where matrix $T = \begin{bmatrix} x_1 & y_1 & 1 \\ x_2 & y_2 & 1 \\ x_3 & y_3 & 1 \end{bmatrix}$.

Example 6: Find the area of the triangle with vertices $(-2,-1)$, $(3,1)$, and $(2,4)$. Assume the units are measured in meters.

Solution: The area of the triangle is found using the formula $A = \left|\dfrac{\det(T)}{2}\right|$.

$T = \begin{bmatrix} -2 & -1 & 1 \\ 3 & 1 & 1 \\ 2 & 4 & 1 \end{bmatrix} \qquad \det(T) = 17$

$A = \left|\dfrac{17}{2}\right| = 8.5\, m^2$

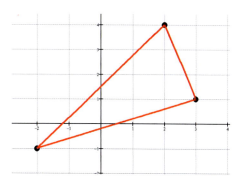

If the determinant of matrix T equals zero, the area would equal zero as well. This would indicate the points do not form a triangle but a straight line. These points are said to be **collinear**.

Example 7: Determine if the following points are collinear.

$(1,1)$, $(3,-5)$, and $(-2,10)$

Solution:

$T = \begin{bmatrix} 1 & 1 & 1 \\ 3 & -5 & 1 \\ -2 & 10 & 1 \end{bmatrix} \qquad \det(T) = 0$

Yes, the points are collinear.

You can use your calculator to find the determinant of a matrix. After the matrix has been entered in the TI-84 calculator, return to the home screen. Press 2ND and x^{-1}. Move the cursor to the MATH submenu and select 1.

FINAL EXAM REVIEW

KEEP IT FRESH

1. Find the factors of the polynomial $P(x) = 6x^4 - 23x^3 + 18x^2 + 12x - 8$.

2. Use the Intermediate Value Theorem to determine if the function has a zero in the interval. $P(x) = x^4 - 2x^2 + 6x - 3$, $[0,1]$

3. Use $P(x) = 2x^3 - 9x^2 + 7x + 6$ to find the following.
a. List all possible rational zeroes for the polynomial using the Rational Zeroes Theorem.
b. Use Descartes' Rule of Signs to determine the possible combinations of real and complex zeroes for the polynomial.
c. Find the zeros of the polynomial.
d. State the end behavior of the polynomial.
e. Graph the function.

4. Write the absolute value function as a piecewise function. Then graph.
$f(x) = |4 - 3x|$

Lesson 13　　　　　　　　　　　　　　　　Practice Exercises

In Exercises 1 – 3, find the determinant of the given matrix. Determine if the matrix is singular or nonsingular.

1. $\begin{bmatrix} 2 & -3 \\ 6 & 5 \end{bmatrix}$
2. $\begin{bmatrix} \frac{2}{3} & 1 \\ -\frac{5}{6} & 2 \end{bmatrix}$
3. $\begin{bmatrix} 1.2 & 0.4 \\ 1.5 & 0.5 \end{bmatrix}$

In Exercises 4 – 6, find the inverse using the formula on page 136.

4. $\begin{bmatrix} 2 & -3 \\ 6 & 5 \end{bmatrix}$
5. $\begin{bmatrix} \frac{2}{3} & 1 \\ -\frac{5}{6} & 2 \end{bmatrix}$
6. $\begin{bmatrix} 4 & -7 \\ 2 & -1 \end{bmatrix}$

In Exercises 7 – 9, find the minor and cofactor for the given entry. Then find the determinant of the given matrix. Determine if the matrix is singular or nonsingular.

7. $A = \begin{bmatrix} 5 & -3 & 2 \\ -9 & 5 & -4 \\ -3 & 1 & -2 \end{bmatrix}$
   Entry a_{11}

8. $B = \begin{bmatrix} 2 & 2 & -1 \\ 1 & -2 & 2 \\ 3 & -1 & 1 \end{bmatrix}$
   Entry b_{32}

9. $C = \begin{bmatrix} -3 & 2 & 4 \\ -1 & -2 & 0 \\ 3 & 1 & 5 \end{bmatrix}$
   Entry c_{22}

In Exercises 10 – 15, solve the system using Cramer's Rule.

10. $\begin{cases} x + 2y = -3 \\ 2x - 3y = 5 \end{cases}$
11. $\begin{cases} 4x - y = -5 \\ 2x + 5y = 13 \end{cases}$
12. $\begin{cases} 5x + 3y = 7 \\ 2x + 5y = 1 \end{cases}$

13. $\begin{cases} 4x - 5y - 6z = 5 \\ 2x - 3y + 3z = 0 \\ x + 2y - 3z = 5 \end{cases}$
14. $\begin{cases} -x + y + 5z = 12 \\ 2x + y = -2 \\ 3x - 2y + z = -8 \end{cases}$
15. $\begin{cases} 3x - z = 8 \\ -y - z = -3 \\ x + 2y + 5z = 10 \end{cases}$

In Exercises 16 – 18, find the area of the triangle with the given vertices. Assume the unit of measurement is meters.

16. $(1,-5); (-2,4); (3,0)$
17. $(-7,3); (-2,1); (5,5)$
18. $(-4,-2); (0,0); (-3,6)$

In Exercises 19 – 21, determine if the points are collinear.

19. $(15,4); (1,-1.6); (10,2)$
20. $(-6,6); (5,-1); (9,-4)$
21. $(-2,-3); (3,4); \left(-\frac{1}{3}, -\frac{2}{3}\right)$

145

Solutions for Practice Exercises **Lesson 13**

1. $\det(A) = 28$; Nonsingular 2. $\det(A) = \dfrac{13}{6}$; Nonsingular 3. $\det(A) = 0$; Singular

4. $\begin{bmatrix} \dfrac{5}{28} & \dfrac{3}{28} \\ -\dfrac{3}{14} & \dfrac{1}{14} \end{bmatrix}$ 5. $\begin{bmatrix} \dfrac{12}{13} & -\dfrac{6}{13} \\ \dfrac{5}{13} & \dfrac{4}{13} \end{bmatrix}$ 6. $\begin{bmatrix} -\dfrac{1}{10} & \dfrac{7}{10} \\ -\dfrac{1}{5} & \dfrac{2}{5} \end{bmatrix}$

7. minor: -6 cofactor: -6 8. minor: 5 cofactor: -5 9. minor: -27 cofactor: -27

$\det(A) = 0$; Singular $\det(A) = 5$; Nonsingular $\det(A) = 60$; Nonsingular

10. $\left(\dfrac{1}{7}, -\dfrac{11}{7}\right)$ 11. $\left(-\dfrac{6}{11}, \dfrac{31}{11}\right)$ 12. $\left(\dfrac{32}{19}, -\dfrac{9}{19}\right)$ 13. $\left(2, 1, -\dfrac{1}{3}\right)$ 14. $\left(-\dfrac{37}{19}, \dfrac{36}{19}, \dfrac{31}{19}\right)$ 15. $\left(\dfrac{14}{5}, \dfrac{13}{5}, \dfrac{2}{5}\right)$

16. $Area = 16.5 m^2$ 17. $Area = 17 m^2$ 18. $Area = 15 m^2$

19. Collinear 20. Not Collinear 21. Collinear

FINAL EXAM REVIEW

KEEP IT FRESH

1. Solve the polynomial inequality. Graph the solution and write the solution in interval notation.

a. $x^3 - x^2 - 4x + 4 > 0$ b. $6x^3 + 19x^2 + 2x \leq 3$

2. Solve the rational inequality. Graph the solution and write the solution in interval notation.

a. $\dfrac{x+2}{x^2 - 9} \leq 0$

b. $\dfrac{x-4}{x+3} \geq \dfrac{x+2}{x-1}$

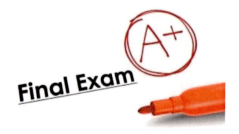

Lesson 14   Partial Fraction Decomposition

In an algebra class, you learned how to add rational expressions. For a review, add the following fractions.

$$\frac{-2}{2x-3} + \frac{3}{x+2} = \frac{-2(x+2)}{(2x-3)(x+2)} + \frac{3(2x-3)}{(2x-3)(x+2)}$$ Find the LCD and get a common denominator.

$$= \frac{-2x-4+6x-9}{(2x-3)(x+2)}$$ Combine like terms.

$$= \frac{4x-13}{(2x-3)(x+2)}$$

In this section, we will learn how to go backwards. In other words, we need to separate the fraction into addition or subtraction. This process is called **partial fraction decomposition**. There are five cases to consider. The cases are not mutually exclusive, which means you can have an expression that includes several cases. The cases depend on the power of the factors in the denominator. After the denominator is factored, determine if the factors are linear or quadratic.

Case 1: Distinct Linear Factors

When the denominator is in factored form, case 1 includes linear factors in the form of $mx+b$ with multiplicity 1. For linear factors, the numerator is a constant. Write the decomposition in the form $\frac{A}{mx+b}$.

Example 1: Decompose $\frac{4x-13}{2x^2+x-6}$ into partial fractions.

Solution: Factor the denominator first. $\frac{4x-13}{2x^2+x-6} = \frac{4x-13}{(2x-3)(x+2)}$

The denominator includes linear factors. Therefore, the numerators will be constants. Use the variables A and B to represent the unknown constants.

$$\frac{4x-13}{(2x-3)(x+2)} = \frac{A}{(2x-3)} + \frac{B}{(x+2)}$$ Multiply both sides of the equation by the LCD. $(2x-3)(x+2)$

$$\cancel{(2x-3)(x+2)} \frac{4x-13}{\cancel{(2x-3)(x+2)}} = \frac{A}{\cancel{(2x-3)}}\cancel{(2x-3)}(x+2) + \frac{B}{\cancel{(x+2)}}(2x-3)\cancel{(x+2)}$$ Cancel common factors.

$$4x-13 = A(x+2) + B(2x-3)$$

There are two values ($x=-2$ and $x=3/2$) that can be used to find A and B. These values are called "convenient values", because they make solving the equation very easy. The convenient

147

values are found by finding the zeros of the LCD. Substitute $x = -2$ into the equation. Notice the variable A will cancel out, which makes solving for B easy.

$$4(-2) - 13 = A(-2 + 2) + B(2(-2) - 3)$$
$$-21 = A(0) - 7B$$
$$3 = B$$

The next convenient value is $x = \dfrac{3}{2}$, which can be substituted into the equation to find A.

$$4\left(\dfrac{3}{2}\right) - 13 = A\left(\dfrac{3}{2} + 2\right) + B\left(2\left(\dfrac{3}{2}\right) - 3\right)$$
$$-7 = \dfrac{7}{2}A + B(0)$$
$$-2 = A$$

Now that we know the values of A and B, we can write the partial fraction decomposition.

$$\dfrac{4x - 13}{(2x - 3)(x + 2)} = \dfrac{-2}{2x - 3} + \dfrac{3}{x + 2}$$

Case 2: Repeated Linear Factors

After the denominator is factored, some linear factors may have multiplicity k. If this occurs, a decomposed fraction is used to represent every multiplicity less than or equal to k. If $(x-1)^4$ is a factor of the denominator, then the partial fraction decomposition would include

$$\dfrac{A}{x-1} + \dfrac{B}{(x-1)^2} + \dfrac{C}{(x-1)^3} + \dfrac{D}{(x-1)^4}.$$

Example 2: Decompose $\dfrac{7x^2 - 29x + 24}{2x^3 - 9x^2 + 12x - 4}$ into partial fractions.

Solution: Factor the denominator first. You will need to use synthetic division.

```
2 | 2  -9   12   -4
         4  -10    4
    2  -5    2    0
```

From the Rational Zeros Theorem, the possible zeros would be $p/q = \pm 1, \pm 2, \pm 4, \pm \frac{1}{2}$. Since the coefficients do not add to 0, $x = 1$ will not be a zero. Try other values in the list until you find a zero remainder.

$(x - 2)(2x^2 - 5x + 2)$
$(x - 2)(2x - 1)(x - 2)$

The factor $(x - 2)$ has multiplicity 2.

$$\frac{7x^2 - 29x + 24}{2x^3 - 9x^2 + 12x - 4} = \frac{7x^2 - 29x + 24}{(2x-1)(x-2)^2}$$

The factors in the denominator are linear factors. Therefore, the numerators will be constants. Use the variables A, B, and C to represent the unknown constants.

$$\frac{7x^2 - 29x + 24}{(2x-1)(x-2)^2} = \frac{A}{(2x-1)} + \frac{B}{(x-2)} + \frac{C}{(x-2)^2}$$

Multiply both sides of the equation by the LCD $(2x-1)(x-2)^2$.

Then cancel common factors to eliminate the fractions.

$$7x^2 - 29x + 24 = \frac{A}{(2x-1)}(2x-1)(x-2)^2 + \frac{B}{(x-2)}(2x-1)(x-2)^2 + \frac{C}{(x-2)^2}(2x-1)(x-2)^2$$

$$7x^2 - 29x + 24 = A(x-2)^2 + B(2x-1)(x-2) + C(2x-1)$$

Use convenient values find the values of A, B, and C. Substitute $x = \frac{1}{2}$ into the equation so variables B and C are multiplied by zero.

$$7\left(\frac{1}{2}\right)^2 - 29\left(\frac{1}{2}\right) + 24 = A\left(\frac{1}{2} - 2\right)^2 + B\left(2\left(\frac{1}{2}\right) - 1\right)\left(\frac{1}{2} - 2\right) + C\left(2\left(\frac{1}{2}\right) - 1\right)$$

$$\frac{45}{4} = \frac{9}{4}A + B(0) + C(0)$$

$$5 = A$$

The next convenient value is $x = 2$, which can be substituted into the equation to find C.

$$7(2)^2 - 29(2) + 24 = A(2-2)^2 + B(2(2)-1)(2-2) + C(2(2)-1)$$

$$-6 = 3C$$

$$-2 = C$$

There are only two convenient values, but there are three variables. What can we do to find B? Now that A and C are known, choose any value for x. One easy value to pick is $x = 0$. Substitute $A = 5$, $C = -2$, and $x = 0$. This will allow us to find B.

$$7(0)^2 - 29(0) + 24 = 5(0-2)^2 + B(2(0)-1)(0-2) - 2(2(0)-1)$$

$$24 = 20 + 2B + 2$$

$$24 = 2B + 22$$

$$2 = 2B$$

$$1 = B$$

$$\frac{7x^2 - 29x + 24}{2x^3 - 9x^2 + 12x - 4} = \frac{5}{(2x-1)} + \frac{1}{(x-2)} - \frac{2}{(x-2)^2}$$

Write the final answer into the partial fraction decomposition.

Case 3: Distinct Quadratic Factors

If a quadratic factor in the form of ax^2+bx+c is prime, the partial fraction decomposition would be in the form $\dfrac{mx+b}{ax^2+bx+c}$. Notice the numerator is linear (one degree less than the denominator).

Example 3: Decompose $\dfrac{11x^2-8x-7}{2x^3-6x^2-x+3}$ into partial fractions.

Solution: Factor the denominator first. You will need to use synthetic division.

$$\frac{11x^2-8x-7}{2x^3-6x^2-x+3} = \frac{11x^2-8x-7}{(x-3)(2x^2-1)}$$

There is one linear factor and one quadratic factor in the denominator. From Case 1, the linear factor will have a constant numerator A. For the quadratic factor, the numerator will be linear. Linear factors are written as $mx+b$, but we will use capital letters B and C instead.

$$\frac{11x^2-8x-7}{(x-3)(2x^2-1)} = \frac{A}{x-3} + \frac{Bx+C}{2x^2-1}$$

Multiply both sides of the equation by $(x-3)(2x^2-1)$.

$$11x^2-8x-7 = A(2x^2-1)+(Bx+C)(x-3)$$

There is only one convenient value here ($x=3$), so we are going to look at a different method to solve for the missing variables. This method includes using a matrix.

$$11x^2-8x-7 = 2Ax^2 - A + Bx^2 - 3Bx + Cx - 3C$$

Remove all parentheses and collect like terms.

$$= 2Ax^2 + Bx^2 - 3Bx + Cx - A - 3C$$

Factor by grouping.

$$= (2A+B)x^2 + (-3B+C)x - A - 3C$$

In order for both sides of the equation to be equal, the coefficients for the variables must be equal. Set up a system of equations, and write the system as an augmented matrix. Use the calculator to find the reduced row-echelon form.

$$2A+B = 11$$
$$-3B+C = -8$$
$$-A -3C = -7$$

$$\begin{bmatrix} 2 & 1 & 0 & | & 11 \\ 0 & -3 & 1 & | & -8 \\ -1 & 0 & -3 & | & -7 \end{bmatrix} \quad \text{RREF} \quad \begin{bmatrix} 1 & 0 & 0 & | & 4 \\ 0 & 1 & 0 & | & 3 \\ 0 & 0 & 1 & | & 1 \end{bmatrix}$$

$A=4 \quad B=3 \quad C=1$

$$\frac{11x^2-8x-7}{2x^3-6x^2-x+3} = \frac{4}{x-3} + \frac{3x+1}{2x^2-1}$$

Case 4: Repeated Quadratic Factors

For quadratic factors with multiplicity k, a decomposed fraction is used to represent every multiplicity less than or equal to k. In other words, $(ax^2+bx+c)^3$ would have 3 decomposed fractions.

$$\frac{}{ax^2+bx+c}+\frac{}{(ax^2+bx+c)^2}+\frac{}{(ax^2+bx+c)^3} \quad \text{All numerators will be linear factors.}$$

> **Example 4:** Decompose $\dfrac{4x^3+5x^2+7x-1}{(x^2+x+1)^2}$ into partial fractions.
>
> **Solution:** Notice the denominator is prime, so we have a quadratic factor that has multiplicity of 2. Each numerator will be in linear form.
>
> $\dfrac{4x^3+5x^2+7x-1}{(x^2+x+1)^2}=\dfrac{Ax+B}{x^2+x+1}+\dfrac{Cx+D}{(x^2+x+1)^2}$ Multiply both sides of the equation by the LCD. The LCD can always be found on the left side of the equation. $(x^2+x+1)^2$
>
> $4x^3+5x^2+7x-1=(Ax+B)(x^2+x+1)+Cx+D$ Remove all parentheses and collect like terms.
>
> $4x^3+5x^2+7x-1=Ax^3+Ax^2+Ax+Bx^2+Bx+B+Cx+D$ Factor by grouping.
>
> $\underline{4}x^3+\underline{5}x^2+\underline{7}x\underline{-1}=\underline{A}x^3+\underline{(A+B)}x^2+\underline{(A+B+C)}x+\underline{B+D}$
>
> $A=4$
> $A+B=5$
> $A+B+C=7$
> $B+D=-1$
>
> This system does not need a matrix. Substitute the known value of A to find B, and so on.
>
> $A=4$
> $B=1$
> $C=2$
> $D=-2$
>
> $\dfrac{4x^3+5x^2+7x-1}{(x^2+x+1)^2}=\dfrac{4x+1}{x^2+x+1}+\dfrac{2x-2}{(x^2+x+1)^2}$

Case 5: Perform Long Division First

If the degree of the numerator is greater than or equal to the degree of the denominator, perform long division first. Then, decompose the $\frac{remainder}{divisor}$ into partial fractions.

Example 5: Decompose $\dfrac{x^3 - 4x^2 - 19x - 35}{x^2 - 7x}$ into partial fractions.

Solution: The degree of the numerator is 3 and the degree of the denominator is 2. Therefore, we must divide first.

$$
\begin{array}{r}
x+3 \\
x^2-7x{\overline{\smash{\big)}\,x^3 - 4x^2 - 19x - 35}} \\
\underline{-(x^3 - 7x^2)} \\
3x^2 - 19x \\
\underline{-(3x^2 - 21x)} \\
2x - 35
\end{array}
$$

$$\frac{x^3 - 4x^2 - 19x - 35}{x^2 - 7x} = x + 3 + \frac{2x - 35}{x^2 - 7x}$$

Decompose $\dfrac{2x - 35}{x^2 - 7x}$ into partial fractions. Factor the denominator first. Notice, both factors are linear. Therefore, we will follow Case 1. The numerators will be constants.

$\dfrac{2x - 35}{x(x-7)} = \dfrac{A}{x} + \dfrac{B}{x-7}$ When solving for A and B, you can multiply each side of the equation by the LCD to remove the fractions.

$x(x-7)\left[\dfrac{2x-35}{x(x-7)}\right] = \left[\dfrac{A}{x} + \dfrac{B}{x-7}\right]x(x-7)$ Multiply both sides by the LCD, $x(x-7)$.

$2x - 35 = A(x - 7) + Bx$

We have two convenient values, $x = 7$ and $x = 0$.

For $x = 7$:

$2(7) - 35 = A(7 - 7) + B(7)$
$-21 = 7B$
$-3 = B$

For $x = 0$:

$2(0) - 35 = A(0 - 7) + B(0)$
$-35 = -7A$
$5 = A$

The partial fraction decomposition is $\dfrac{5}{x} + \dfrac{-3}{x-7}$.

The final answer includes the quotient and the partial fraction decomposition.

$$\frac{x^3 - 4x^2 - 19x - 35}{x^2 - 7x} = x + 3 + \frac{5}{x} - \frac{3}{x-7}$$

You will use the skills learned in the section in calculus. For example, you will be required to integrate a problem that includes a rational function.

$$\int \frac{x^3 - 4x^2 - 19x - 35}{x^2 - 7x} dx$$

In order to perform the calculus operation, a student must first write the function using long division and partial fraction decomposition. Do not forget these algebra techniques. You will see them again!

$$\int x + 3 + \frac{5}{x} - \frac{3}{x-7} dx$$

FINAL EXAM REVIEW

KEEP IT FRESH

1. Use properties of logarithms to write the expression as a single logarithm. Simplify if possible.

   a. $2\log x + 3\log y - \frac{1}{2}\log z$

   b. $\ln(x^2 - 4) - \ln(x+2) + \ln x$

2. Use properties of logarithms to expand the logarithm as sums/differences.

   a. $\log \sqrt{100x}$ b. $\ln \frac{e}{y^3}$

3. Graph the following functions.

   a. $f(x) = 2^{x+3} - 5$ b. $g(x) = -\ln(x+3)$

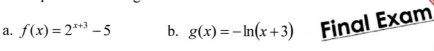

153

Lesson 14 Practice Exercises

In Exercises 1 – 22, decompose each rational expression into partial fractions.

1. $\dfrac{7x-23}{x^2-7x+12}$

2. $\dfrac{2x+46}{x^2-4x-21}$

3. $\dfrac{-x-33}{x^2+3x-18}$

4. $\dfrac{9x+2}{x^2+x-6}$

5. $\dfrac{5}{6x^2-x-1}$

6. $\dfrac{2x-4}{x^2+2x+1}$

7. $\dfrac{x^2-9x+21}{x^3-6x^2+12x-8}$

8. $\dfrac{-8x^2-38x-41}{x^3+9x^2+27x+27}$

9. $\dfrac{3x^2-22x+152}{(x-7)^2(4x+1)}$

10. $\dfrac{4x^3+9x^2+x}{(2x+3)^2(x^2+x+1)}$

11. $\dfrac{-x^3+5x-3}{x^4+5x^2+4}$

12. $\dfrac{3x^3+4x^2+20x-6}{(x^2+x+2)(x^2-x+5)}$

13. $\dfrac{9x^2-16x+32}{x^3-8}$

14. $\dfrac{x^2+28x-33}{x^3+27}$

15. $\dfrac{x^3-12x^2-5x-40}{x^4+2x^2-3}$

16. $\dfrac{4x^3+7x^2+9x+3}{x^4+4x^2+4}$

17. $\dfrac{2x^3+5x^2-31x-12}{2x^2-7x-4}$

18. $\dfrac{6x^3-10x^2+29x-17}{(x^2+5x+1)(2x^2+3)}$

19. $\dfrac{6x^3+9x^2+7x-3}{(3x^2+1)^2}$

20. $\dfrac{4x^4-18x^3+31x^2-26x+25}{(x^2+1)^2(x-4)}$

21. $\dfrac{2x^3-15x^2+21x-21}{x^2-7x}$

22. $\dfrac{3x^3+7x^2-4x+1}{x^2}$

23. $\dfrac{x^4-3x^3-2x^2-4x-31}{x^2-3x-4}$

24. $\dfrac{x^5+3x^4+9x^3+15x^2+21x+4}{x^3+4x}$

25. $\dfrac{x^3-7x^2+10x-58}{(x^2+9)^2}$

26. $\dfrac{x^5+6x^4+8x^3+16x^2+15x+10}{(x+1)^2(x^2+1)^2}$

27. $\dfrac{2x^3-4x-8}{x^4-x^3+4x^2-4x}$

Solutions for Practice Exercises **Lesson 14**

1. $\dfrac{2}{x-3}+\dfrac{5}{x-4}$

2. $\dfrac{6}{x-7}-\dfrac{4}{x+3}$

3. $\dfrac{3}{x+6}-\dfrac{4}{x-3}$

4. $\dfrac{4}{x-2}+\dfrac{5}{x+3}$

5. $\dfrac{2}{2x-1}-\dfrac{3}{3x+1}$

6. $\dfrac{2}{x+1}-\dfrac{6}{(x+1)^2}$

7. $\dfrac{7}{(x-2)^3}-\dfrac{5}{(x-2)^2}+\dfrac{1}{x-2}$

8. $\dfrac{1}{(x+3)^3}+\dfrac{10}{(x+3)^2}-\dfrac{8}{x+3}$

9. $\dfrac{5}{(x-7)^2}+\dfrac{3}{4x+1}$

10. $\dfrac{3}{(2x+3)^2}+\dfrac{2}{2x+3}-\dfrac{1}{x^2+x+1}$

11. $\dfrac{2x-1}{x^2+1}+\dfrac{-3x+1}{x^2+4}$

12. $\dfrac{x-4}{x^2+x+2}+\dfrac{2x+7}{x^2-x+5}$

13. $\dfrac{3}{x-2}+\dfrac{6x-10}{x^2+2x+4}$

14. $\dfrac{-4}{x+3}+\dfrac{5x+1}{x^2-3x+9}$

15. $\dfrac{6}{x+1}-\dfrac{7}{x-1}+\dfrac{2x+1}{x^2+3}$

16. $\dfrac{x-11}{(x^2+2)^2}+\dfrac{4x+7}{x^2+2}$

17. $x+6-\dfrac{1}{2x+1}+\dfrac{8}{x-4}$

18. $\dfrac{3x-7}{x^2+5x+1}+\dfrac{4}{2x^2+3}$

19. $\dfrac{5x-6}{(3x^2+1)^2}+\dfrac{2x+3}{3x^2+1}$

20. $\dfrac{2x}{(x^2+1)^2}+\dfrac{3x-6}{x^2+1}+\dfrac{1}{x-4}$

21. $2x-1+\dfrac{3}{x}+\dfrac{11}{x-7}$

22. $3x+7+\dfrac{1}{x^2}-\dfrac{4}{x}$

23. $x^2+2+\dfrac{5}{x+1}-\dfrac{3}{x-4}$

24. $x^2+3x+5+\dfrac{1}{x}+\dfrac{2x+1}{x^2+4}$

25. $\dfrac{x-7}{x^2+9}+\dfrac{x+5}{(x^2+9)^2}$

26. $\dfrac{1}{x+1}+\dfrac{2}{(x+1)^2}+\dfrac{3}{x^2+1}+\dfrac{4}{(x^2+1)^2}$

27. $\dfrac{2}{x}-\dfrac{2}{x-1}+\dfrac{2x+4}{x^2+4}$

Cumulative Review 3   Lessons 10 – 14

1. Solve the system of equations using Gaussian or Gauss-Jordan elimination method.

   a. $\begin{cases} 2x+y=1 \\ 3x+2y=-2 \end{cases}$

   b. $\begin{cases} 2x+y=1 \\ 3x-6y=4 \end{cases}$

   c. $\begin{cases} 4x-8y=12 \\ -x+2y=-3 \end{cases}$

   d. $\begin{cases} 2x-2y+z=-3 \\ x-2y+3z=-1 \\ x-y+2z=0 \end{cases}$

   e. $\begin{cases} 3x+2y+2z=3 \\ x+2y-z=5 \\ 2x-4y+z=0 \end{cases}$

   f. $\begin{cases} 2x+3y+7z=-7 \\ 4x+5y+z=1 \\ x+y-3z=4 \end{cases}$

2. Given: $A = \begin{bmatrix} 1 & 2 \\ 4 & 3 \end{bmatrix}$ $B = \begin{bmatrix} -3 & 4 \\ 6 & 7 \end{bmatrix}$ $C = \begin{bmatrix} 2 & -3 & 4 \\ 4 & 0 & -1 \\ 5 & -2 & 0 \end{bmatrix}$

   $D = \begin{bmatrix} 2 & -3 & 0 \\ 4 & 0 & -2 \end{bmatrix}$ $E = \begin{bmatrix} 1 & 2 & -1 \\ -2 & 0 & 1 \\ 1 & -1 & 0 \end{bmatrix}$

   a. $2A+3B$ b. $3C-2E$ c. $C+2D$ d. DC e. AB
   f. CE g. DA h. A^{-1} i. C^{-1} j. E^{-1}
   k. $\det(A)$ l. $\det(C)$ m. $\det(E)$

3. Write the system of equations as a matrix equation. Then solve the equation using the inverse of the coefficient matrix.

   a. $\begin{cases} 2x-3y=7 \\ 4x+y=-7 \end{cases}$

   b. $\begin{cases} -x+4y=-5 \\ x-6y=5 \end{cases}$

   c. $\begin{cases} x+2y+3z=-1 \\ 2x-3y+4z=2 \\ -3x+5y-6z=4 \end{cases}$

4. Solve using Cramer's Rule.

   a. $\begin{cases} 3x+2y=7 \\ 2x+3y=-2 \end{cases}$

   b. $\begin{cases} 2x+3y=-1 \\ 3x+6y=-0.5 \end{cases}$

   c. $\begin{cases} 3x+2y+2z=1 \\ 5x-y-6z=3 \\ 2x+3y+3z=4 \end{cases}$

   d. $\begin{cases} x-2y-3z=4 \\ 3x-2z=8 \\ 2x+y+4z=13 \end{cases}$

5. Find the area of a triangle with vertices:
   a. (–2,2), (1,5), and (6,–1).   b. (1,0), (2,2), (4,3)

6. Determine if the points are collinear.
   a. (0,1), (2,2), (4,3)   b. (–2,–2), (1,1), (7,5)

7. Decompose into partial fractions.

   a. $\dfrac{13x+46}{12x^2-11x-15}$

   b. $\dfrac{5x^3+6x^2+5x}{(x^2-1)(x+1)^3}$

   c. $\dfrac{11x^2-39x+16}{(x^2+4)(x-8)}$

   d. $\dfrac{x^2-10x+13}{(x^2-5x+6)(x-1)}$

   e. $\dfrac{2x^3+3x^2-11x-10}{x^2+2x-3}$

Solutions to Review

1a. $(4,-7)$ 1b. $\left(\dfrac{2}{3},-\dfrac{1}{3}\right)$ 1c. $(2y+3, y)$ 1d. $(0,2,1)$ 1e. $\left(2,\dfrac{1}{2},-2\right)$ 1f. $(16z+19, -13z-15, z)$

2a. $\begin{bmatrix} -7 & 16 \\ 26 & 27 \end{bmatrix}$ 2b. $\begin{bmatrix} 4 & -13 & 14 \\ 16 & 0 & -5 \\ 13 & -4 & 0 \end{bmatrix}$ 2c. cannot add 2d. $\begin{bmatrix} -8 & -6 & 11 \\ -2 & -8 & 16 \end{bmatrix}$

2e. $\begin{bmatrix} 9 & 18 \\ 6 & 37 \end{bmatrix}$ 2f. $\begin{bmatrix} 12 & 0 & -5 \\ 3 & 9 & -4 \\ 9 & 10 & -7 \end{bmatrix}$ 2g. cannot multiply 2h. $\begin{bmatrix} -\dfrac{3}{5} & \dfrac{2}{5} \\ \dfrac{4}{5} & -\dfrac{1}{5} \end{bmatrix}$

2i. $\begin{bmatrix} \dfrac{2}{21} & \dfrac{8}{21} & -\dfrac{1}{7} \\ \dfrac{5}{21} & \dfrac{20}{21} & -\dfrac{6}{7} \\ \dfrac{8}{21} & \dfrac{11}{21} & -\dfrac{4}{7} \end{bmatrix}$ 2j. $\begin{bmatrix} 1 & 1 & 2 \\ 1 & 1 & 1 \\ 2 & 3 & 4 \end{bmatrix}$ 2k. -5 2l. -21 2m. 1

3a. $(-1,-3)$ 3b. $(5,0)$ 3c. $(124,14,-51)$

4a. $(5,-4)$ 4b. $\left(-\dfrac{3}{2},\dfrac{2}{3}\right)$ 4c. $(-1,4,-2)$ 4d. $(4,-3,2)$

5a. 16.5 5b. 3/2 6a. yes 6b. no

7a. $\dfrac{-5}{4x+3}+\dfrac{7}{3x-5}$ 7b. $\dfrac{1}{x-1}-\dfrac{1}{x+1}+\dfrac{3}{(x+1)^2}-\dfrac{3}{(x+1)^3}+\dfrac{2}{(x+1)^4}$

7c. $\dfrac{5x+1}{x^2+4}+\dfrac{6}{x-8}$ 7d. $\dfrac{-4}{x-3}+\dfrac{3}{x-2}+\dfrac{2}{x-1}$ 7e. $2x-1+\dfrac{1}{x+3}-\dfrac{4}{x-1}$

Lesson 15 Factor Completely

For any polynomial, there are two steps to factor completely.

Step 1: Factor out the greatest common factor. (GCF)

Example: $\underline{8x^4 - 12x^3}$ ← *Note: The GCF for the variable is always the smallest exponent.*

$4x^3(2x - 3)$ *Also, division is used to determine the polynomial in the parentheses.* $\frac{8x^4}{4x^3} = 2x$ $\frac{12x^3}{4x^3} = 3$

Step 2: Count the number of terms.

For 2 terms, look for the following options.

Difference of Squares	$a^2 - b^2 = (a-b)(a+b)$
Difference of Cubes	$a^3 - b^3 = (a-b)(a^2 + ab + b^2)$
Sum of Cubes	$a^3 + b^3 = (a+b)(a^2 - ab + b^2)$
Note: The sum of squares is prime.	$a^2 + b^2$

For 3 terms, use trinomial techniques.

Perfect Square Trinomial $a^2 \pm 2ab + b^2 = (a \pm b)^2$

FOIL – Trial and Error

For 4 terms, factor by grouping or synthetic division.

Example 1: Factor completely. $(x^3 + 1)^3(2x) - (x^2 - 1)(3)(x^3 + 1)^2 3x^2$

Solution: You may find it easier to find the GCF by organizing the terms first.

$2x(x^3+1)^3 - 9x^2(x^2-1)(x^3+1)^2$ *Divide each term by the GCF* $x(x^3+1)^2$. *Subtract exponents.*

$x(x^3+1)^2[2(x^3+1) - 9x(x^2-1)]$ $\frac{2x(x^3+1)^3}{x(x^3+1)^2} = 2(x^3+1)$ $\frac{9x^2(x^2-1)(x^3+1)^2}{x(x^3+1)^2} = 9x(x^2-1)$

$x(x^3+1)^2[2x^3 + 2 - 9x^3 + 9x]$

$x(x^3+1)^2(-7x^3 + 9x + 2)$ *If the first term is negative, you can factor out* -1.

$-x(x^3+1)^2(7x^3 - 9x - 2)$ *The trinomial* $7x^3 - 9x - 2$ *can be factored using synthetic division.*

$-x(x+1)(x^3+1)^2(7x^2 - 7x - 2)$ *Note: The binomial* (x^3+1) *is the sum of cubes. However, we will not use the formula for factoring because it has multiplicity of 2.*

In this lesson, we need to expand the concept to factor expressions with negative and fractional exponents. Even though these expressions are not polynomials, we can use the same two step method to factor completely.

> **Example 2:** Factor completely. Write your answer with positive exponents.
>
> a. $24x^{-1/2} + 16x^{1/2}$
>
> b. $(x-1)^{-2} - 3(x-1)^{-1}$
>
> c. $6(x-2)^{1/2}(3x+1)^2 - x(3x+1)^3(x-2)^{-1/2}$
>
> d. $(3x+2)^{2/3}\left(\dfrac{1}{2}\right)(x^2-16)(2x) + (x^2-16)^2\left(\dfrac{1}{3}\right)(3x+2)^{-1/3}(9)$

Solution:

a. $24x^{-1/2} + 16x^{1/2}$

First, find the GCF $8x^{-1/2}$. Always use the smallest exponent for the GCF.

To determine the expression in parentheses, we will use division.

$24x^{-1/2} + 16x^{1/2}$
$8x^{-1/2}(3 + 2x)$

$\dfrac{24x^{-1/2}}{8x^{-1/2}} = 3$ Since the exponents are the same, the variable x cancels completely.

$\dfrac{16x^{1/2}}{8x^{-1/2}} = 2x$ When dividing, the exponents are subtracted. $x^{1/2-(-1/2)} = x^1$

The binomial is prime. After you move the negative exponent to the denominator, the final answer is $\dfrac{8(2x+3)}{x^{1/2}}$.

---

b. $(x-1)^{-2} - 3(x-1)^{-1}$

The GCF is $(x-1)^{-2}$.

$(x-1)^{-2}[1 - 3(x-1)]$

$(x-1)^{-2}[1 - 3x + 3]$

$(x-1)^{-2}(-3x + 4)$

$\dfrac{-(3x-4)}{(x-1)^2}$

$\dfrac{(x-1)^{-2}}{(x-1)^{-2}} = 1$

$\dfrac{3(x-1)^{-1}}{(x-1)^{-2}} = 3(x-1)^{-1-(-2)} = 3(x-1)$

The binomial in the parentheses is prime. If the first term is negative, you can factor out a −1.

To write the answer with positive exponents, move negative exponents to the denominator.

c. $6(x-2)^{1/2}(3x+1)^2 - x(3x+1)^3(x-2)^{-1/2}$ The GCF includes $(x-2)^{-1/2}(3x+1)^2$.

$(x-2)^{-1/2}(3x+1)^2[6(x-2)^{1/2-(-1/2)} - x(3x+1)^{3-2}]$

$(x-2)^{-1/2}(3x+1)^2[6(x-2) - x(3x+1)]$

$(x-2)^{-1/2}(3x+1)^2[6x - 12 - 3x^2 - x]$

$(x-2)^{-1/2}(3x+1)^2(-3x^2 + 5x - 12)$

$\dfrac{(3x+1)^2(-3x^2+5x-12)}{(x-2)^{1/2}}$ or $\dfrac{-(3x+1)^2(3x^2-5x+12)}{(x-2)^{1/2}}$

It is important to always collect like terms in the bracket. Use distributive property to remove parentheses. Check to see if the polynomial can be factored.

$-3x^2 + 5x - 12 = -(3x^2 - 5x + 12)$

To determine if the trinomial factors, look at the "AC" method.

AC	SUM of B
3(12)	−5
36	
-1(-36)	Both signs are negative.
-2(-18)	None of the factors add
-3(-12)	to −5. Therefore, the
-4(-9)	trinomial is prime.
-6(-6)	

---

d. Organize the terms first. $(3x+2)^{2/3}\left(\dfrac{1}{2}\right)(x^2-16)(2x) + (x^2-16)^2\left(\dfrac{1}{3}\right)(3x+2)^{-1/3}(9)$

$\left(\dfrac{1}{2}\right)(2x)(3x+2)^{2/3}(x^2-16) + \left(\dfrac{1}{3}\right)(9)(x^2-16)^2(3x+2)^{-1/3}$

$x(3x+2)^{2/3}(x^2-16) + 3(x^2-16)^2(3x+2)^{-1/3}$

$(3x+2)^{-1/3}(x^2-16)[x(3x+2)^{2/3-(-1/3)} + 3(x^2-16)^{2-1}]$

$(3x+2)^{-1/3}(x^2-16)[3x^2 + 2x + 3x^2 - 48]$

$(3x+2)^{-1/3}(x^2-16)(6x^2 + 2x - 48)$

$2(3x+2)^{-1/3}(x^2-16)(3x^2 + x - 24)$

$2(3x+2)^{-1/3}(x+4)(x-4)(3x-8)(x+3)$

$\dfrac{2(x+4)(x-4)(3x-8)(x+3)}{(3x+2)^{1/3}}$

Note: the binomial $(x^2 - 16)$ is the difference of two squares. We will not factor this until the end. Find the GCF first and divide to determine the polynomial in the bracket. Distribute to remove parentheses and collect like terms. Now, determine if the trinomial can be factored.

There is a common factor of 2 that is placed at the very beginning of the answer.

$6x^2 + 2x - 48$
$2(3x^2 + x - 24)$
$2(3x-8)(x+3)$

Don't forget to move factors with negative exponents to the denominator.

The last type of problem that we need to be able to do in this section has one additional step. After factoring as in Examples 1 and 2, we will cancel any common factors with the denominator. Don't make the mistake of trying to cancel first. It is important to notice that we factor the numerator, then cancel any common factors with the denominator.

Example 3: Simplify. Write your answer with positive exponents.

a. $\dfrac{2x(1-4x)^3 + 6x(1-4x)^2}{4(1-4x)^6}$

b. $\dfrac{\dfrac{1}{2}(x-3)^{1/2} + \dfrac{3}{2}(x-3)^{-1/2}}{\left[(x-3)^{1/3}\right]^3}$

Solution:

a. Start by factoring the numerator. This is the same process as Examples 1 and 2.

$$\frac{2x(1-4x)^2[(1-4x)+3]}{4(1-4x)^6}$$

Divide each term by the GCF $2x(1-4x)^2$. Subtract exponents.

$$\frac{2x(1-4x)^3}{2x(1-4x)^2} = (1-4x)^{3-2} \qquad \frac{6x(1-4x)^2}{2x(1-4x)^2} = 3$$

$$\frac{2x(1-4x)^2[4-4x]}{4(1-4x)^6}$$

The binomial $4-4x$ can be factored. Notice the GCF 4 is placed at the beginning.

$$\frac{4 \cdot 2x(1-4x)^2(1-x)}{4(1-4x)^6}$$

Now, cancel common factors in the numerator and denominator.

$$\frac{2x(1-x)}{(1-4x)^4} \quad \text{or} \quad \frac{-2x(x-1)}{(1-4x)^4}$$

---

b. Factor the numerator first. For fractions $\frac{1}{2}$ and $\frac{3}{2}$, the GCF of the numerator is 1 and the GCF of the denominator is 2. Therefore, the GCF of the fraction is $\frac{1}{2}$.

$$\frac{\frac{1}{2}(x-3)^{1/2} + \frac{3}{2}(x-3)^{-1/2}}{\left[(x-3)^{1/3}\right]^3}$$

Multiply

To determine the polynomial in the bracket, divide by the GCF.

$$\frac{\frac{1}{2}(x-3)^{-1/2}[x-3+3]}{(x-3)}$$

$$\frac{\frac{1}{2}(x-3)^{1/2}}{\frac{1}{2}(x-3)^{-1/2}} = (x-3)^{1/2-(-1/2)} \qquad \frac{\frac{3}{2}(x-3)^{-1/2}}{\frac{1}{2}(x-3)^{-1/2}} = \frac{\frac{3}{2}}{\frac{1}{2}} = 3$$

$$\frac{2}{2} \cdot \frac{\frac{1}{2}x}{(x-3)(x-3)^{1/2}}$$

Move factors with negative exponents to the denominator.

The factors in the denominator combine by adding exponents.

$$\frac{x}{2(x-3)^{3/2}}$$

Fix the complex fraction by multiplying the numerator and denominator by 2.

When will you need to use such difficult factoring skills? In calculus, your answer will result in binomials that must be completely factored. The factored form will allow you to solve the expression equal to 0.

Solve $\dfrac{2x(1-4x)^3 + 6x(1-4x)^2}{4(1-4x)^6} = 0$. Using the factored form in Example 3a, we can see the solutions are $\{0,1\}$ and $x = \dfrac{1}{4}$ is an undefined value.

Lesson 15 Practice Exercises

In Exercises 1 – 16, factor the expression completely. Write your answer with positive exponents.

1. $(x+2)^2 - 5(x+2)$

2. $(x-1)^2 - 2(x-1)$

3. $6x^2(1-2x)^4 - 24x^3(1-2x)^3$

4. $6(3x-5)(2x-3)^2 + 4(3x-5)^2(2x-3)$

5. $2(x-3)(4x+7)^2 + 8(x-3)^2(4x+7)$

6. $5x^4(9-x)^4 - 4x^5(9-x)^3$

7. $3x^4(x-7)^2 + 4x^3(x-7)^3$

8. $2(x+1)^3(x^2-5)^2 + 4x(x+1)^4(x^2-5)$

9. $6(x^2-4)^{\frac{1}{2}}(2x+1)^2 + x(2x+1)^3(x^2-4)^{-\frac{1}{2}}$

10. $8(3x+2)^{\frac{1}{3}}(4x-5) + (4x-5)^2(3x+2)^{-\frac{2}{3}}$

11. $(3x+1)^6(2x-5)^{-\frac{1}{2}} + (2x-5)^{\frac{1}{2}}(3x+1)^5$

12. $(x^2+9)^4(x+6)^{-\frac{4}{3}} + 2x(x+6)^{-\frac{1}{3}}(x^2+9)^3$

13. $(x-12)^3(2)(x^2-9)(2x) + (x^2-9)^2(3)(x-12)^2(4)$

14. $(3x+2)^{\frac{2}{3}}\left(\frac{1}{2}\right)(x^2-16)(2x) + (x^2-16)^2\left(\frac{1}{3}\right)(3x+2)^{-\frac{1}{3}}(9)$

15. $(x+2)^{-\frac{1}{4}}\left(\frac{1}{2}\right)(x-6)^3(5x) + (x-6)^2\left(\frac{1}{3}\right)(x+2)^{\frac{3}{4}}(7)$

16. $(x-2)^{-5}\left(\frac{1}{4}\right)(x+2)^3(3x) + (x-2)^{-6}\left(\frac{1}{3}\right)(x+2)^4(2)$

In Exercises 17 – 26, simplify the expressions. Write your answer with positive exponents.

17. $\dfrac{6x^3(x^2+2)^2 - 2x(x^2+2)^3}{x^4}$

18. $\dfrac{4x^4(x^2+3)^2 - 3x^2(x^2+3)^3}{x^6}$

19. $\dfrac{2x(1-3x)^3 + 9x^2(1-3x)^2}{(1-3x)^6}$

20. $\dfrac{2x(2x+3)^4 - 8x^2(2x+3)^3}{(2x+3)^9}$

21. $\dfrac{(6x+1)^3(27x^2+2) - (9x^3+2x)(18)(6x+1)^2}{(6x+1)^6}$

22. $\dfrac{(x^2-1)^4(2x) - x^2(4)(x^2-1)^3(2x)}{(x^2-1)^8}$

23. $\dfrac{(x^2-5)^4(3x^2) - x^3(4)(x^2-5)^3(2x)}{\left[(x^2-5)^4\right]^2}$

24. $\dfrac{(x^2+4)^{\frac{1}{3}} + 2x^2(x^2+4)^{-\frac{2}{3}}}{\left[(x^2+4)^{\frac{1}{3}}\right]^2}$

25. $\dfrac{\frac{1}{2}(x+5)^{\frac{1}{2}} + \frac{3}{2}(x+5)^{\frac{-1}{2}}}{\left[(x+5)^{\frac{1}{2}}\right]^3}$

26. $\dfrac{10x(x^2+4)^4(x-6)^4 + 4(x^2+4)^5(x-6)^3}{(x-6)^5}$

Solutions for Practice Exercises — Lesson 15

1. $(x+2)(x-3)$
2. $(x-1)(x-3)$
3. $6x^2(1-2x)^3(1-6x)$

4. $2(3x-5)(2x-3)(12x-19)$
5. $2(x-3)(4x+7)(8x-5)$
6. $9x^4(9-x)^3(5-x)$

7. $7x^3(x-7)^2(x-4)$
8. $2(x+1)^3(x^2-5)(3x+5)(x-1)$
9. $\dfrac{(2x+1)^2(8x^2+x-24)}{[(x+2)(x-2)]^{1/2}}$

10. $\dfrac{(4x-5)(28x+11)}{(3x+2)^{2/3}}$
11. $\dfrac{(3x+1)^5(5x-4)}{(2x-5)^{1/2}}$
12. $\dfrac{3(x^2+9)^3(x+1)(x+3)}{(x+6)^{4/3}}$

13. $4(x-12)^2(x+3)(x-3)(2x+3)(2x-9)$
14. $\dfrac{2(x+4)(x-4)(3x-8)(x+3)}{(3x+2)^{1/3}}$

15. $\dfrac{(x-6)^2(3x-14)(5x-2)}{6(x+2)^{1/4}}$
16. $\dfrac{(x+2)^3(9x^2-10x+16)}{12(x-2)^6}$

17. $\dfrac{4(x^2+2)^2(x+1)(x-1)}{x^3}$
18. $\dfrac{(x^2+3)^2(x+3)(x-3)}{x^4}$
19. $\dfrac{x(3x+2)}{(1-3x)^4}$

20. $\dfrac{2x(3-2x)}{(2x+3)^6}$
21. $\dfrac{27x^2-24x+2}{(6x+1)^4}$
22. $\dfrac{-2x(3x^2+1)}{(x^2-1)^5}$

23. $\dfrac{-5x^2(x^2+3)}{(x^2-5)^5}$
24. $\dfrac{3x^2+4}{(x^2+4)^{4/3}}$
25. $\dfrac{x+8}{2(x+5)^2}$

26. $\dfrac{2(x^2+4)^4(7x-2)(x-4)}{(x-6)^2}$

Lesson 16 — Circle and Ellipse

Two right circular cones intersected by a plane form **conic sections**. In this lesson, we will study the circle and the ellipse.

Circle   Ellipse

A **circle** is the set of all points in a plane that are at a fixed distance (radius) from a fixed point (center) in the plane.

From any point on the circle, the distance to the center is constant. This constant is called the radius r. Using the distance formula, we can find the standard form of the circle.

$d = \sqrt{(x_2 - x_1)^2 + (y_2 - y_1)^2}$

$r = \sqrt{(x-h)^2 + (y-k)^2}$

$(x-h)^2 + (y-k)^2 = r^2$

(h,k) is the center of the circle, and r is the radius.

Example 1: Find the standard form of the circle. Determine the center and the radius.

$x^2 + y^2 - 6x + 4y - 3 = 0$

Solution: First, you need to organize the variables. Then complete the square.

$x^2 - 6x + \underline{} + y^2 + 4y + \underline{} = 3 + \underline{} + \underline{}$

$x^2 - 6x + \underline{\;9\;} + y^2 + 4y + \underline{\;4\;} = 3 + \underline{\;9\;} + \underline{\;4\;}$

$\left(\dfrac{b}{2}\right)^2$

$(x-3)^2 + (y+2)^2 = 16$

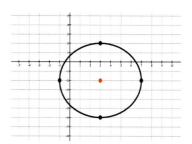

Center is $(3,-2)$ and the radius is $r = 4$

An **ellipse** is the set of all points in a plane, the sum of whose distances from two fixed points is constant. The two fixed points are called **foci** (plural form of focus). The midpoint between the foci is the center of the ellipse.

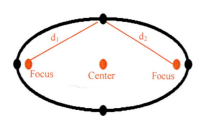

The sum of the distances is constant.

$d_1 + d_2 = k$

The standard form for an ellipse is $\dfrac{(x-h)^2}{a^2}+\dfrac{(y-k)^2}{b^2}=1$. The center of the ellipse is (h,k).

The variable a represents the horizontal distance from the center. The variable b represents the vertical distance from the center.

Horizontal Major Axis
$a > b$

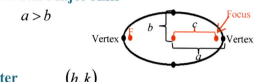

Center	(h,k)
Vertices	$(h \pm a, k)$
Foci	$(h \pm c, k)$
Eccentricity	$e = \dfrac{c}{a}$
Length of Major Axis	$2a$

Vertical Major Axis
$b > a$

Center	(h,k)
Vertices	$(h, k \pm b)$
Foci	$(h, k \pm c)$
Eccentricity	$e = \dfrac{c}{b}$
Length of Major Axis	$2b$

Notice that variable c represents the distance from the center to the foci. The variable c is not in the equation, so how do you find the distance? Let's look specifically at a horizontal major axis with center $(0,0)$. The sum of the distances from $(0,b)$ to the foci will equal the sum of the distances from $(a,0)$ to the foci.

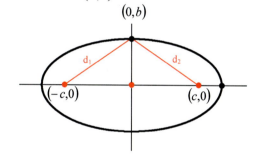 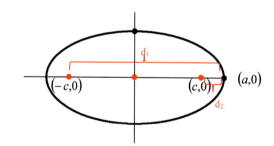

$$d_1 + d_2$$
$$\sqrt{(0-(-c))^2 + (b-0)^2} + \sqrt{(0-c)^2 + (b-0)^2}$$
$$\sqrt{c^2 + b^2} + \sqrt{c^2 + b^2}$$
$$2\sqrt{c^2 + b^2}$$

$$d_1 + d_2$$
$$\sqrt{(a-(-c))^2 + (0-0)^2} + \sqrt{(a-c)^2 + (0-0)^2}$$
$$\sqrt{(a+c)^2} + \sqrt{(a-c)^2}$$
$$a + c + a - c$$
$$2a$$

Set the two sums equal to find the relationship between variables a, b, and c.

$$2\sqrt{c^2 + b^2} = 2a \qquad \text{\color{red}Divide by 2.}$$
$$\sqrt{c^2 + b^2} = a \qquad \text{\color{red}Square both sides.}$$
$$c^2 + b^2 = a^2 \qquad \text{\color{red}Subtract } b^2.$$
Horizontal Major Axis $\qquad c^2 = a^2 - b^2$

Using a similar proof for the vertical major axis, the relationship is the equation $c^2 = b^2 - a^2$.

As the foci approach the center, the ellipse becomes circular. In fact, if the value of $c = 0$, we have a circle instead of an ellipse. To compare the "roundness" of an ellipse, we use the **eccentricity**. Because $c < a$ and $c < b$, the eccentricity will always be a value between 0 and 1 ($0 < e < 1$). For an ellipse with $e \approx 0$, the graph is almost a circle. For an ellipse with $e \approx 1$, the graph is long and thin.

> **Example 2:** Use the given information to write the equation of the ellipse in standard form.
>
> **a.** Foci $(5,-2)$ and $(5,4)$ & Vertices $(5,-4)$ and $(5,6)$
>
> **b.** Eccentricity $e = \dfrac{2}{5}$ and Major Axis is on the x-axis with length 20 and Center $(0,0)$

Solution:

a. To find the equation, we first need to determine if the major axis is horizontal or vertical. Draw a sketch of the information given. Find the center and determine the distances from the center to the points.

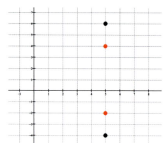

From the points, we can determine the major axis is vertical. Find the midpoint of the vertices to determine the center.

$$\left(\frac{x_1+x_2}{2}, \frac{y_1+y_2}{2}\right) = \left(\frac{5+5}{2}, \frac{-4+6}{2}\right) = (5,1)$$

The distance between the center and a vertex point is b. $b = 5$
The distance between the center and a focus point is c. $c = 3$

Since the only variable needed to find the equation is a, we can use the formula $b^2 - a^2 = c^2$.

$5^2 - a^2 = 3^2$
$a = 4$

Write the equation in standard form. $\dfrac{(x-5)^2}{16} + \dfrac{(y-1)^2}{25} = 1$

---

b. The center is given. Since the major axis is on the *x*-axis, we know the ellipse has a horizontal major axis. Therefore, the length $2a = 20$ gives the value $a = 10$. Use the formula for the eccentricity to solve for c.   Now, find the value of b.   Write the equation.

$e = \dfrac{c}{a}$ $\dfrac{c}{10} = \dfrac{2}{5}$ $a^2 - b^2 = c^2$
       $5c = 20$ $10^2 - b^2 = 4^2$ $\dfrac{x^2}{100} + \dfrac{y^2}{84} = 1$
       $c = 4$ $b = 2\sqrt{21}$

If the eccentricity is given, do not assume the numerator is the value of c and the denominator is the value of a. The fraction value of e will be in reduced form.

Example 3: Write the equation of the ellipse in standard form. Identify the center, vertices, foci, length of major, length minor axis, and eccentricity. Then sketch the graph.

a. $3(x-4)^2 + 4(y+1)^2 = 12$
b. $4x^2 + y^2 - 24x - 8y + 48 = 0$

Solution:

a. $3(x-4)^2 + 4(y+1)^2 = 12$

$\dfrac{3(x-4)^2}{12} + \dfrac{4(y+1)^2}{12} = \dfrac{12}{12}$

$\dfrac{(x-4)^2}{4} + \dfrac{(y+1)^2}{3} = 1$

$h = 4$
$k = -1$
$a = 2$
$b = \sqrt{3}$
$c = 1$

Center $(4,-1)$

Vertices
$(4 \pm 2, -1)$
$(2,-1)$ and $(6,-1)$

Foci
$(4 \pm 1, -1)$
$(3,-1)$ and $(5,-1)$

Length of major axis
$2a = 4$

Length of minor axis
$2b = 2\sqrt{3}$

Eccentricity $\quad e = \dfrac{c}{a} = \dfrac{1}{2}$

b. $4x^2 + y^2 - 24x - 8y + 48 = 0$

$4x^2 - 24x + \underline{} + y^2 - 8y + \underline{} = -48 + \underline{} + \underline{}$

$4(x^2 - 6x + \underline{9}) + y^2 - 8y + \underline{16} = -48 + \underline{36} + \underline{16}$

$4(x-3)^2 + (y-4)^2 = 4$

$\dfrac{(x-3)^2}{1} + \dfrac{(y-4)^2}{4} = 1$

$h = 3$
$k = 4$
$a = 1$
$b = 2$ ← Since $b > a$, the major axis is vertical.
$c = \sqrt{3}$

Center $(3,4)$

Vertices
$(3,2)$ and $(3,6)$

Length of major axis
$2b = 4$

Eccentricity
$e = \dfrac{c}{b} = \dfrac{\sqrt{3}}{2}$

Foci
$(3, 4 \pm \sqrt{3})$

Length of minor axis
$2a = 2$

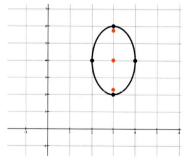

There are many applications for the ellipse. When you exercise on an elliptical machine, you are using an ellipse. The orbits of the planets around the sun can be modeled using an ellipse. In front of the White House, there is a park nicknamed the Ellipse for its shape. The base of the Colosseum in Rome has an elliptical shape. Who's in for a field trip? The foci of the ellipse have reflective properties that are very useful in science, such as sound waves and light rays. There is an elliptical dome shape building called a whispering gallery that allows whispers to be heard clearly. See exercise 26.

Example 4: A semi-elliptical archway over a one-way road has a height of 10 feet and a width of 40 feet. A truck has a width of 10 feet and a height of 9 feet. Will the truck clear the opening of the archway?

Solution: Draw a picture of the archway on the x-y axis. Label the values that are given. For application problems, you can put the ellipse anywhere on the graph. To make calculations easier, use the origin for the center.

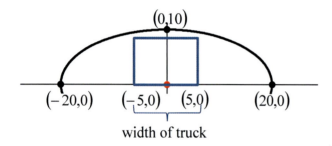

Determine the equation needed. We have a horizontal major axis with center $(0,0)$, horizontal distance $a = 20$, and vertical distance $b = 10$.

$$\frac{x^2}{400} + \frac{y^2}{100} = 1$$

Determine the y value when $x = 5$.

$$\frac{5^2}{400} + \frac{y^2}{100} = 1$$

$$\frac{y^2}{100} = 1 - \frac{25}{400}$$

$$(100)\frac{y^2}{100} = \frac{15}{16}(100)$$

$$y^2 = \frac{375}{4}$$

$$y \approx 9.682$$

Will your truck that is 9 feet high clear the opening of the archway? Yes. Barely! The height of the arch is approximately 9.682 feet when $x = 5$. Hold your breath as you drive under the archway with about 8 inches of clearance.

Lesson 16 Practice Exercises

In Exercises 1 – 6, find the center of the ellipse.

1. Vertices $(0,4)$ and $(0,-4)$

2. Vertices $(-1,3)$ and $(-5,3)$

3. Foci $(5,0)$ and $(-5,0)$

4. Foci $(-2,8)$ and $(-2,-2)$

5. $\dfrac{(x-4)^2}{9} + \dfrac{(y+1)^2}{25} = 1$

6. $\dfrac{(x+8)^2}{16} + \dfrac{(y-2)^2}{4} = 1$

In Exercises 7 – 10, find the length of the major axis.

7. Vertices $(1,-2)$ and $(1,4)$

8. Vertices $(7,0)$ and $(-2,0)$

9. $\dfrac{x^2}{16} + \dfrac{y^2}{36} = 1$

10. $\dfrac{(x-4)^2}{100} + \dfrac{(y-1)^2}{36} = 1$

In Exercises 11 – 14, use the given information to write the equation of the ellipse in standard form.

11. Foci $(0,\pm 6)$ and minor axis length 4

12. Foci $(7,2)$ and $(-1,2)$ and major axis length 10

13. Foci $(\pm 5,0)$ and Vertices $(\pm 6,0)$

14. Vertices $(-3,-3)$ and $(-3,7)$ and minor axis length 4

In Exercises 15 – 24, write the equation of the ellipse in standard form. Identify the center, vertices, foci, length of major, length minor axis, and eccentricity. Then sketch the graph.

15. $16(x-2)^2 + 9(y-1)^2 = 144$

16. $4(x+3)^2 + 25(y+1)^2 = 100$

17. $4x^2 + y^2 = 4$

18. $(x-5)^2 + 9y^2 = 36$

19. $25x^2 + 4y^2 + 150x - 16y + 141 = 0$

20. $9x^2 + 25y^2 - 54x - 100y - 44 = 0$

21. $64x^2 + 100y^2 - 6400 = 0$

22. $16x^2 + 7y^2 - 128x + 42y + 207 = 0$

23. $6x^2 + 9y^2 + 24x + 36y + 6 = 0$

24. $5x^2 + 2y^2 - 50x - 12y + 93 = 0$

25. The area of an ellipse is given by the formula $A = \pi ab$. In Calculus, you will prove this formula is true using integration. Find the area of the ellipse given by the following equations. Assume the units are measured in feet.

a. $\dfrac{x^2}{25} + \dfrac{y^2}{36} = 1$

b. $\dfrac{(x-2)^2}{6} + \dfrac{(y+3)^2}{9} = 1$

c. $9x^2 + 16y^2 - 54x - 64y + 1 = 0$

26. A room 88 feet long is constructed to be a whispering gallery. The room has an elliptical ceiling. If the maximum height of the ceiling is 22 feet, determine how far from the center of the room the whispering dishes should be placed so that a whisper can be heard. (Whispering dishes are placed at the foci of an ellipse.)

Solutions for Practice Exercises Lesson 16

1. $(0,0)$ 2. $(-3,3)$ 3. $(0,0)$ 4. $(-2,3)$ 5. $(4,-1)$ 6. $(-8,2)$

7. 6   8. 9   9. 12   10. 20   11. $\dfrac{x^2}{4}+\dfrac{y^2}{40}=1$

12. $\dfrac{(x-3)^2}{25}+\dfrac{(y-2)^2}{9}=1$ 13. $\dfrac{x^2}{36}+\dfrac{y^2}{11}=1$ 14. $\dfrac{(x+3)^2}{4}+\dfrac{(y-2)^2}{25}=1$

15. $\dfrac{(x-2)^2}{9}+\dfrac{(y-1)^2}{16}=1$ 16. $\dfrac{(x+3)^2}{25}+\dfrac{(y+1)^2}{4}=1$

center $(2,1)$ vertices $(2,5)\,\&\,(2,-3)$ center $(-3,-1)$ vertices $(2,-1)\,\&\,(-8,-1)$

foci $(2,1\pm\sqrt{7})$ major 8 minor 6   foci $(-3\pm\sqrt{21},-1)$ major 10 minor 4

$e=\dfrac{\sqrt{7}}{4}$ $e=\dfrac{\sqrt{21}}{5}$

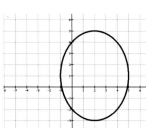   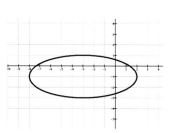

17. $\dfrac{x^2}{1}+\dfrac{y^2}{4}=1$ 18. $\dfrac{(x-5)^2}{36}+\dfrac{y^2}{4}=1$

center $(0,0)$ vertices $(0,2)\,\&\,(0,-2)$ center $(5,0)$ vertices $(11,0)\,\&\,(-1,0)$

foci $(0,\pm\sqrt{3})$ major 4 minor 2   foci $(5\pm 4\sqrt{2},0)$ major 12 minor 4

$e=\dfrac{\sqrt{3}}{2}$ $e=\dfrac{2\sqrt{2}}{3}$

19. $\dfrac{(x+3)^2}{4}+\dfrac{(y-2)^2}{25}=1$ 20. $\dfrac{(x-3)^2}{25}+\dfrac{(y-2)^2}{9}=1$

center $(-3,2)$ vertices $(-3,7)\,\&\,(-3,-3)$ center $(3,2)$ vertices $(8,2)\,\&\,(-2,2)$

foci $(-3,2\pm\sqrt{21})$ major 10 minor 4   foci $(7,2)\,\&\,(-1,2)$ major 10 minor 6

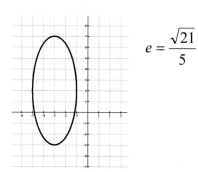 $e = \dfrac{\sqrt{21}}{5}$

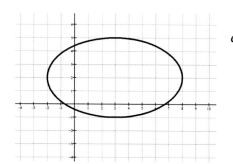 $e = \dfrac{4}{5}$

21. $\dfrac{x^2}{100} + \dfrac{y^2}{64} = 1$

center $(0,0)$ vertices $(\pm 10, 0)$

foci $(\pm 6, 0)$ major 20 minor 16 $e = \dfrac{2}{5}$

22. $\dfrac{(x-4)^2}{7} + \dfrac{(y+3)^2}{16} = 1$

center $(4,-3)$ vertices $(4,1) \& (4,-7)$

foci $(4,0) \& (4,-6)$ major 8 minor $2\sqrt{7}$ $e = \dfrac{3}{4}$

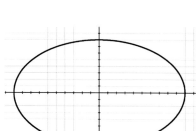

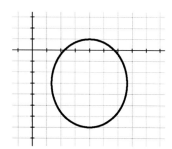

23. $\dfrac{(x+2)^2}{9} + \dfrac{(y+2)^2}{6} = 1$

center $(-2,-2)$ vertices $(1,-2) \& (-5,-2)$

foci $(-2 \pm \sqrt{3}, -2)$ major 6 minor $2\sqrt{6}$

24. $\dfrac{(x-5)^2}{10} + \dfrac{(y-3)^2}{25} = 1$

center $(5,3)$ vertices $(5,8) \& (5,-2)$

foci $(5, 3 \pm \sqrt{15})$ major 10 minor $2\sqrt{10}$

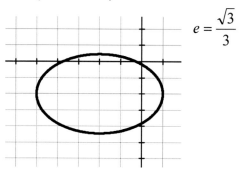 $e = \dfrac{\sqrt{3}}{3}$

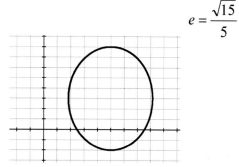 $e = \dfrac{\sqrt{15}}{5}$

25a. $30\pi \ ft^2$ 25b. $3\pi\sqrt{6} \ ft^2$ 25c. $12\pi \ ft^2$ 26. $c = 22\sqrt{3} ft \approx 38.1 ft$

Lesson 17 — Hyperbola

A **hyperbola** is the set of all points in a plane for which the absolute value of the difference from two fixed points is constant. The two fixed points are the foci.

Hyperbola

The segment connecting the two vertices is called the **transverse axis**. The center of the hyperbola is the midpoint of the transverse axis.

Horizontal Transverse Axis

$$\frac{(x-h)^2}{a^2} - \frac{(y-k)^2}{b^2} = 1$$

$a^2 + b^2 = c^2$

Center (h, k)

Vertices $(h \pm a, k)$

Foci $(h \pm c, k)$

Asymptotes $y - k = \pm \frac{b}{a}(x - h)$

Length of Transverse Axis $2a$

Eccentricity $e = \frac{c}{a}$

Vertical Transverse Axis

$$\frac{(y-k)^2}{b^2} - \frac{(x-h)^2}{a^2} = 1$$

$a^2 + b^2 = c^2$

Center (h, k)

Vertices $(h, k \pm b)$

Foci $(h, k \pm c)$

Asymptotes $y - k = \pm \frac{b}{a}(x - h)$

Length of Transverse Axis $2b$

Eccentricity $e = \frac{c}{b}$

Eccentricity measures the "wideness" of the hyperbola. Since $c > a$ and $c > b$, the eccentricity will always be greater than 1. If e is close to 1, the foci points are close to the center, making the hyperbola narrow. As e increases, the distance from the foci to the center increases, making the hyperbola wider. If $e = 1$, the graph is a parabola.

The line segment passing through the center of the hyperbola and is perpendicular to the transverse axis is called the **conjugate axis**.

Example 1: Use the given information to write the equation of the hyperbola in standard form.

a. Vertices $(\pm 3, 0)$ and Foci $(\pm 4, 0)$

b. Vertices $(-1, \pm 4)$ and Asymptotes $y = \pm \frac{1}{2}(x + 1)$

Solution:

a. Draw a sketch of the given information to determine the hyperbola has a horizontal transverse axis. Find the midpoint of the transverse axis to find the center. The distance between the center and one vertex point is a. The distance between the center and one focus point is c. Use the equation $a^2 + b^2 = c^2$ to find the value of b.

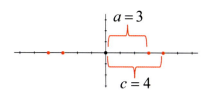

Midpoint of Vertices

$\left(\dfrac{3+(-3)}{2}, \dfrac{0+0}{2}\right)$

$(0,0)$

$h = 0, \ k = 0$

$a^2 + b^2 = c^2$

$3^2 + b^2 = 4^2$

$b^2 = 16 - 9$

$b^2 = 7$

Equation:
$$\dfrac{x^2}{9} - \dfrac{y^2}{7} = 1$$

---

b. One assumption would be to label $a = 2$ and $b = 1$ using the slope of the asymptote. This would be a mistake given the location of the vertices. Remember, slope is a reduced fraction. Therefore, the slope $\pm\dfrac{1}{2}$ could have many different values for a and b. One thing we can gather from the asymptotes is the center of the hyperbola, which can also be found using the midpoint of the vertices.

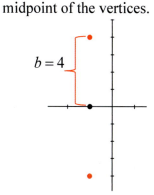

Asymptotes $\quad y = \pm\dfrac{1}{2}(x+1)$

$k = 0, \ h = -1$

Slope $\quad \dfrac{b}{a} = \dfrac{1}{2}$

$\dfrac{4}{a} = \dfrac{1}{2}\quad$ cross multiply

$a = 8$

Since the transverse axis is vertical, the equation for the hyperbola must have the y variable first.

$$\dfrac{y^2}{16} - \dfrac{(x+1)^2}{64} = 1$$

The asymptotes are very important for determining the branches of the hyperbola. You can always use the point-slope formula $y - k = \pm\dfrac{b}{a}(x - h)$ to graph the asymptotes. However, there is another useful method involving a **central rectangle** that will help you graph the asymptotes. After you plot the center, find the vertical and horizontal points using the values or a and b.

Sketch vertical and horizontal line segments through these points as shown below. Draw the diagonals of the central rectangle. These diagonals have slope $\pm \frac{b}{a}$. If you extend the diagonals, you obtain the asymptotes of the hyperbola.

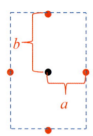
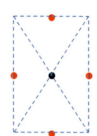
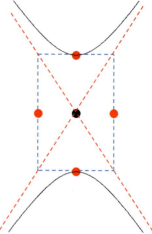

Example 2: **Write the equation of the hyperbola in standard form. Identify the center, vertices, foci, asymptotes, and eccentricity. Then sketch the graph.**

a. $4x^2 - 25y^2 + 16x + 50y - 109 = 0$ b. $x^2 - y^2 - 6x + 8y - 3 = 0$

Solution:

a. Organize the variables to complete the square. Remember: When completing the square, the leading coefficient must be 1. Factor out 4 and –25 first. Then add $\left(\frac{b}{2}\right)^2$ to both sides.

$$4x^2 + 16x \underline{\quad} - 25y^2 + 50y \underline{\quad} = 109 + \underline{\quad} + \underline{\quad}$$
$$4(x^2 + 4x + \underline{\ 4\ }) - 25(y^2 - 2y + \underline{\ 1\ }) = 109 + \underline{\ 16\ } + \underline{-25}$$
$$4(x+2)^2 - 25(y-1)^2 = 100$$
$$\frac{4(x+2)^2}{100} - \frac{25(y-1)^2}{100} = 1$$
$$\frac{(x+2)^2}{25} - \frac{(y-1)^2}{4} = 1$$

$h = -2$

$k = 1$

$a = 5$

$b = 2$

$c = \sqrt{29}$

Center $(-2, 1)$

Vertices $(-2 \pm 5, 1)$

$(-7, 1)$ and $(3, 1)$

Foci $(-2 \pm \sqrt{29}, 1)$

Asymptotes

$y - 1 = \pm \frac{2}{5}(x+2)$

Eccentricity

$e = \dfrac{c}{a} = \dfrac{\sqrt{29}}{5}$

Since *x* is first, the hyperbola has a horizontal transverse axis.

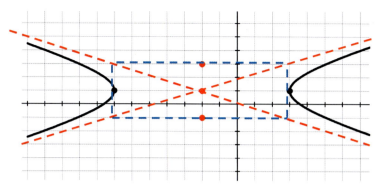

b. $x^2 - y^2 - 6x + 8y - 3 = 0$

$x^2 - 6x \underline{} - y^2 + 8y \underline{} = 3 + \underline{} + \underline{}$
$x^2 - 6x + \underline{9} - (y^2 - 8y + \underline{16}) = 3 + \underline{9} + \underline{-16}$
$(x-3)^2 - (y-4)^2 = -4$
$\dfrac{(x-3)^2}{-4} - \dfrac{(y-4)^2}{-4} = 1$
$\dfrac{(y-4)^2}{4} - \dfrac{(x-3)^2}{4} = 1$

$h = 3$
$k = 4$
$a = 2$
$b = 2$
$c = 2\sqrt{2}$

Center $(3,4)$

Vertices $(3, 4 \pm 2)$

$(3,2)$ and $(3,6)$

Foci $(3, 4 \pm 2\sqrt{2})$

Asymptotes
$y - 4 = \pm(x - 3)$

Eccentricity

$e = \dfrac{c}{b} = \dfrac{2\sqrt{2}}{2} = \sqrt{2}$

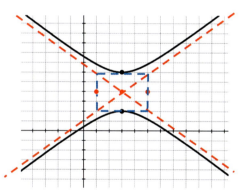

Example 3: Some telescopes use a hyperbolic mirror, which has the property that a light ray directed at the focus will be reflected to the other focus. Find the equation that represents the hyperbolic curve of the mirror. At which point on the mirror will light from the point be reflected to the other focus?

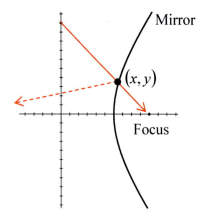

Solution: From the picture, we can determine the center of the hyperbolic mirror is $(0,0)$. The distance from the center to the focus is $c = 10$ and the distance to the vertex is $a = 6$. Using the equation, $a^2 + b^2 = c^2$, we can find the variable b.
$$6^2 + b^2 = 10^2$$
$$b = 8$$

Since the transverse axis is horizontal, we have the equation $\dfrac{(x-h)^2}{a^2} - \dfrac{(y-k)^2}{b^2} = 1$. Substituting the known values, the equation that represents the hyperbolic curve of the mirror is $\dfrac{x^2}{36} - \dfrac{y^2}{64} = 1$.

To determine the point (x, y) on the mirror that light from the point $(0,10)$ will be reflected to the other focus, we need to find the intersection of the line and hyperbola. The light ray represents a linear equation with y-intercept $(0,10)$ and slope $m = -1$, $y = -x + 10$. Using a system of equations, we can find the intersection with the substitution method.

$$\begin{cases} \dfrac{x^2}{36} - \dfrac{y^2}{64} = 1 \\ y = -x + 10 \end{cases}$$

Replace y in the hyperbolic equation with $-x + 10$.

$$576\left(\dfrac{x^2}{36} - \dfrac{(-x+10)^2}{64}\right) = (1)576 \quad \text{Clear fractions by multiplying by the LCD.}$$

$$16x^2 - 9(x^2 - 20x + 100) = 576 \quad \text{Collect like terms.}$$
$$16x^2 - 9x^2 + 180x - 900 = 576$$
$$7x^2 + 180x - 1476 = 0 \quad \text{Solve the equation using the quadratic formula.}$$

$$x = \dfrac{-180 \pm \sqrt{180^2 - 4(7)(-1476)}}{2(7)}$$

$$x \approx 6.5378 \text{ and } x \approx -32.2521$$

From the picture, we need only the positive x value. Find the y value using the linear equation.
$y = -(6.5378) + 10$
$y \approx 3.4622$

From the point $(0,10)$, light will be reflected to the other focus at the point $(6.5378, 3.4622)$ on the hyperbolic mirror.

There are other applications using the hyperbolic equation. Light from a lamp forms a hyperbolic shadow on a wall. Some comets pass through our solar system using a hyperbolic pattern. The cross section of a nuclear cooling tower forms a hyperbolic equation. The LORAN (long range navigation) system uses a hyperbolic pattern to determine the location of a ship.

Example 4: A nuclear cooling tower is 300 feet tall and the distance from the top of the tower to the center of the hyperbola is half the distance from the base of the tower to the center of the hyperbola. Find the diameter of the top and the base of the tower. The cross section of the tower is represented by the equation $\dfrac{x^2}{64^2} - \dfrac{y^2}{100^2} = 1$.

Solution: Sketch a picture of the tower using the equation given. The hyperbola will have a horizontal transverse axis with center at the origin, $a = 64$ and $b = 100$.

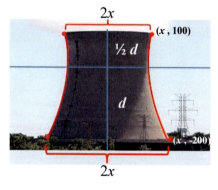

Let d represent the distance from the base to the center. Since the top to the center is half the distance from the base to the center, $\dfrac{1}{2}d$ represents the distance from the top to the center.

Therefore, $\dfrac{1}{2}d + d = 300$. Solve for d. The distance to the base is 200 ft and the distance to the top is 100 ft. To find the diameter of the top and base of the tower we will need to find the value of x when the $y = 100$ and $y = -200$.

For $y = 100$:

$$\dfrac{x^2}{64^2} - \dfrac{100^2}{100^2} = 1$$

$$\dfrac{x^2}{64^2} - 1 = 1$$

$$\dfrac{x^2}{64^2} = 2$$

$$x^2 = 2(64^2)$$

$$\sqrt{x^2} = \sqrt{2(64^2)}$$

$$x = 64\sqrt{2}$$

For $y = -200$:

$$\dfrac{x^2}{64^2} - \dfrac{(-200)^2}{100^2} = 1$$

$$\dfrac{x^2}{64^2} - 4 = 1$$

$$\dfrac{x^2}{64^2} = 5$$

$$x^2 = 5(64^2)$$

$$\sqrt{x^2} = \sqrt{5(64^2)}$$

$$x = 64\sqrt{5}$$

The value of x is the radius. Multiply by 2 to find the diameter. The diameter is $2(64\sqrt{2}) = 128\sqrt{2} \approx 181$ feet at the top of the tower and $2(64\sqrt{5}) = 128\sqrt{5} \approx 286$ feet at the bottom of the tower.

Lesson 17　　　　　　　　　　　　Practice Exercises

In Exercises 1 – 6, find the center of the hyperbola.

1. Vertices $(0,7)$ and $(0,-1)$

2. Vertices $(-3,2)$ and $(-7,2)$

3. Foci $(1,0)$ and $(-1,0)$

4. Foci $(-2,8)$ and $(-2,-2)$

5. Asymptotes $y-8=\pm\frac{1}{2}(x+1)$

6. $\frac{x^2}{16}-\frac{y^2}{9}=1$

In Exercises 7 – 10, find the length of the transverse axis.

7. Vertices $(-3,2)$ and $(-3,10)$

8. Vertices $(4,0)$ and $(-2,0)$

9. $\frac{(x-3)^2}{25}-\frac{(y-1)^2}{36}=1$

10. $y^2-(x+2)^2=4$

In Exercises 11 – 14, use the given information to write the equation of the hyperbola in standard form.

11. Foci $(0,\pm 3)$ and conjugate axis length 2

12. Foci $(1,-4)$ and $(-9,-4)$ and transverse axis length 6

13. Vertices $(\pm 6,0)$ and Foci $(\pm 8,0)$

14. Asymptotes $y=\pm\frac{2}{3}x$　Vertices $(0,\pm 4)$

In Exercises 15 – 24, write the equation of the hyperbola in standard form. Identify the center, vertices, foci, asymptotes, and eccentricity. Then sketch the graph.

15. $4x^2-16y^2=64$

16. $4(y+3)^2-9(x+1)^2=36$

17. $(y-1)^2-(x-2)^2=9$

18. $16(x-5)^2-25y^2=400$

19. $9y^2-x^2+54y+4x+68=0$

20. $7x^2-9y^2-14x+72y-200=0$

21. $16x^2-12y^2+144=0$

22. $5y^2-2x^2+20y+8x+62=0$

23. $4x^2-y^2+40x-4y+60=0$

24. $20y^2-16x^2+80y-128x-496=0$

25. An architect designs two apartment buildings that are shaped and positioned like the branches of the hyperbola whose equation is $196y^2-100x^2=19,600$, where x and y are in feet. Write the equation in standard form and determine the eccentricity of the branches. How far apart are the apartment buildings at their closet point?

Solutions for Practice Exercises **Lesson 17**

1. $(0,3)$ 2. $(-5,2)$ 3. $(0,0)$ 4. $(-2,3)$ 5. $(-1,8)$

6. $(0,0)$ 7. 8   8. 6   9. 10   10. 4

11. $\dfrac{y^2}{8} - x^2 = 1$ 12. $\dfrac{(x+4)^2}{9} - \dfrac{(y+4)^2}{16} = 1$ 13. $\dfrac{x^2}{36} - \dfrac{y^2}{28} = 1$ 14. $\dfrac{y^2}{16} - \dfrac{x^2}{36} = 1$

15. $\dfrac{x^2}{16} - \dfrac{y^2}{4} = 1$ 16. $\dfrac{(y+3)^2}{9} - \dfrac{(x+1)^2}{4} = 1$

center $(0,0)$ vertices $(4,0) \& (-4,0)$ center $(-1,-3)$ vertices $(-1,0) \& (-1,-6)$

foci $(\pm 2\sqrt{5}, 0)$ asymptotes $y = \pm\dfrac{1}{2}x$ foci $(-1, -3 \pm \sqrt{13})$ asymptotes $y + 3 = \pm\dfrac{3}{2}(x+1)$

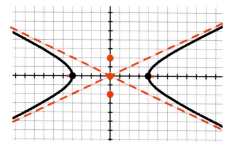 $e = \dfrac{\sqrt{5}}{2}$ 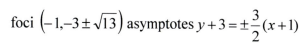 $e = \dfrac{\sqrt{13}}{3}$

17. $\dfrac{(y-1)^2}{9} - \dfrac{(x-2)^2}{9} = 1$ 18. $\dfrac{(x-5)^2}{25} - \dfrac{y^2}{16} = 1$

center $(2,1)$ vertices $(2,4) \& (2,-2)$ center $(5,0)$ vertices $(10,0) \& (0,0)$

foci $(2, 1 \pm 3\sqrt{2})$ asymptotes $y - 1 = \pm(x-2)$ foci $(5 \pm \sqrt{41}, 0)$ asymptotes $y = \pm\dfrac{4}{5}(x-5)$

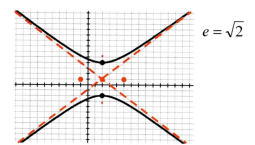 $e = \sqrt{2}$ 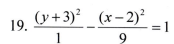 $e = \dfrac{\sqrt{41}}{5}$

19. $\dfrac{(y+3)^2}{1} - \dfrac{(x-2)^2}{9} = 1$ 20. $\dfrac{(x-1)^2}{9} - \dfrac{(y-4)^2}{7} = 1$

center $(2,-3)$ vertices $(2,-2) \& (2,-4)$ center $(1,4)$ vertices $(4,4) \& (-2,4)$

foci $(2,-3\pm\sqrt{10})$ asymptotes $y+3=\pm\frac{1}{3}(x-2)$

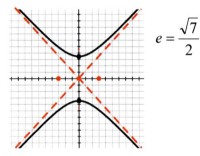

$e=\sqrt{10}$

foci $(-3,4) \& (5,4)$ asymptotes $y-4=\pm\frac{\sqrt{7}}{3}(x-1)$

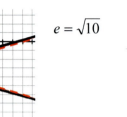

$e=\frac{4}{3}$

21. $\frac{y^2}{12}-\frac{x^2}{9}=1$

22. $\frac{(x-2)^2}{25}-\frac{(y+2)^2}{10}=1$

center $(0,0)$ vertices $(0,\pm 2\sqrt{3})$

center $(2,-2)$ vertices $(7,-2) \& (-3,-2)$

foci $(0,\pm\sqrt{21})$ asymptotes $y=\pm\frac{2\sqrt{3}}{3}x$

foci $(2\pm\sqrt{35},-2)$ asymptotes $y+2=\pm\frac{\sqrt{10}}{5}(x-2)$

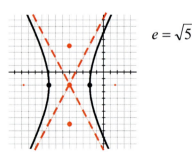

$e=\frac{\sqrt{7}}{2}$

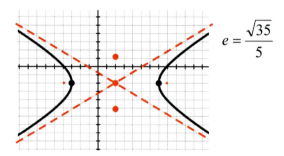

$e=\frac{\sqrt{35}}{5}$

23. $\frac{(x+5)^2}{9}-\frac{(y+2)^2}{36}=1$

24. $\frac{(y+2)^2}{16}-\frac{(x+4)^2}{20}=1$

center $(-5,-2)$ vertices $(-2,-2) \& (-8,-2)$

center $(-4,-2)$ vertices $(-4,2) \& (-4,-6)$

foci $(-5\pm 3\sqrt{5},-2)$ asymptotes $y+2=\pm 2(x+5)$

foci $(-4,4) \& (-4,-8)$ asymptotes $y+2=\pm\frac{2\sqrt{5}}{5}(x+4)$

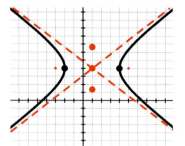

$e=\sqrt{5}$

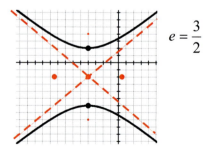

$e=\frac{3}{2}$

25. $\frac{y^2}{100}-\frac{x^2}{196}=1$; $e=\frac{\sqrt{74}}{5}$; 20 ft

Lesson 18 — Parabola

A **parabola** is the set of all points in a plane equidistant from a fixed line (called a **directrix**) to a fixed point (called the **focus**) not on the line.

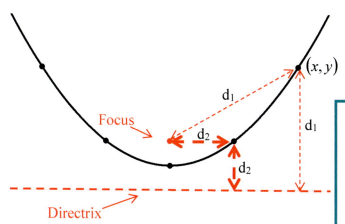

Parabola

The distance from any point on the parabola to the focus will equal the distance from the point on the parabola to the directrix.

There are three components for the parabola that must be determined before graphing.

  Vertex, Focus, and Directrix

All of these components can be found using the standard form of the equation.

The standard form of a parabola is $(x-h)^2 = 4p(y-k)$ or $(y-k)^2 = 4p(x-h)$.

For $(x-h)^2 = 4p(y-k)$, the parabola will open up if $p > 0$ and down if $p < 0$. When the variable x is squared, the resulting parabola is a function. The graph will pass the vertical line test. The following formulas can be used to find the three components.

Vertex: (h, k) Focus: $(h, k+p)$ Directrix: $y = k - p$

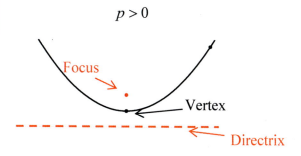

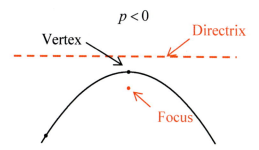

The directrix is below the parabola.   The directrix is above the parabola.

For $(y-k)^2 = 4p(x-h)$, the parabola will open right if $p > 0$ and left if $p < 0$. When the variable y is squared, the resulting parabola is not a function. The graph will not pass the vertical line test. The following formulas can be used to find the three components.

Vertex: (h, k) Focus: $(h+p, k)$ Directrix: $x = h - p$

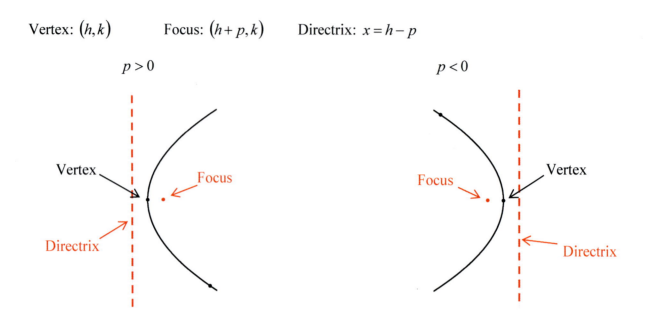

The directrix is left of the parabola. The directrix is right of the parabola.

Notice the focus point is always inside the curve of the parabola. In other words, the parabola will always open toward the focus point.

A line segment that passes through the focus of a parabola and has endpoints on the parabola is called a **focal chord**. The specific focal chord perpendicular to the axis of symmetry is the **latus rectum**. The length of the latus rectum is 4p. To plot the endpoints of the latus rectum, count a distance of 2p on each side of the focus point.

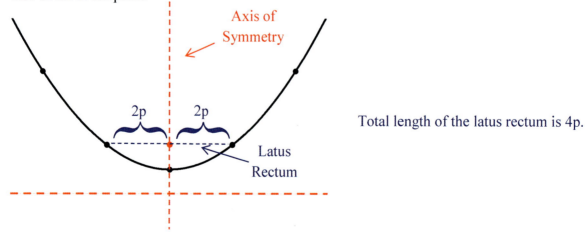

Total length of the latus rectum is 4p.

Example 1: Find the standard form of the conic section $x^2 - 4x - 8y - 20 = 0$. Determine the components and graph.

Solution: Complete the square to put the equation in standard form.

$x^2 - 4x \underline{} = 8y + 20 \underline{}$ The number in front of the x^2 term must be 1.

$x^2 - 4x + 4 = 8y + 20 + 4$ Then add $\left(\dfrac{b}{2}\right)^2$ to both sides. $\left(\dfrac{b}{2}\right)^2 = \left(\dfrac{4}{2}\right)^2 = 4$

$(x-2)^2 = 8y + 24$

$(x-2)^2 = 8(y+3)$ Factor both sides.

$h = 2 \quad 4p = 8 \quad k = -3$ Now, find the variables for all the components.
$ p = 2$

Vertex: $(h, k) = (2, -3)$ Focus: $(h, k+p) = (2, -1)$ Directrix: $y = -5$

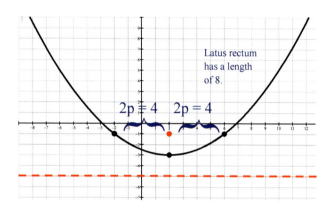

Latus rectum has a length of 8.

$2p = 4 \quad 2p = 4$

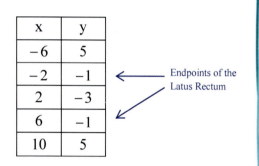

x	y
-6	5
-2	-1
2	-3
6	-1
10	5

← Endpoints of the Latus Rectum

When graphing a parabola, you need to plot at least 5 points. The vertex and the endpoints of the latus rectum will give you 3. The other two points can be the x or y intercepts. You can also use the axis of symmetry to locate more points. Make a table of values if needed.

Example 2: Find the standard form of the conic section. Then determine the components and graph. $y^2 - 2y + 12x + 13 = 0$

Solution: Complete the square to put the equation in standard form.

$y^2 - 2y \underline{} = -12x - 13 \underline{}$ The number in front of the y^2 term must be 1.

$y^2 - 2y + 1 = -12x - 13 + 1$ Then add $\left(\dfrac{b}{2}\right)^2$ to both sides. $\left(\dfrac{b}{2}\right)^2 = \left(\dfrac{-2}{2}\right)^2 = 1$

$(y-1)^2 = -12x - 12$ Factor both sides.

$(y-1)^2 = -12(x+1)$ Now, find the variables for all the components.

$k = 1 \quad 4p = -12 \quad h = -1$
$\quad\quad\quad p = -3$

Vertex: $(h, k) = (-1, 1)$ Focus: $(h+p, k) = (-4, 1)$ Directrix: $x = h - p = 2$

This parabola will not cross the y-axis, but it does cross the x-axis. Evaluate the equation for y = 0 to find the x-intercept.

$y^2 + 2y + 12x + 13 = 0$
$0^2 + 2(0) + 12x + 13 = 0$
$\quad\quad\quad\quad 12x + 13 = 0$
$\quad\quad\quad\quad\quad\quad x = -\dfrac{13}{12}$

x	y
-4	7
$-\dfrac{13}{12}$	2
-1	1
$-\dfrac{13}{12}$	0
-4	-5

Latus rectum has a length of 12.

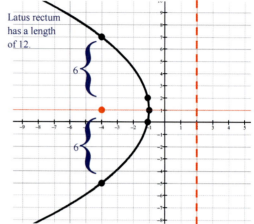

Now, use symmetry to find the other point $\left(-\dfrac{13}{12}, 2\right)$.

Example 3: Find the standard form of the conic section $2x^2 + 8x + 4y + 3 = 0$. Then determine the components.

Solution: Complete the square to put the equation in standard form.

$2x^2 + 8x \underline{\quad} = -4y - 3 \underline{\quad}$ The number in front of the x^2 term must be 1. Factor out 2.

$2(x^2 + 4x \underline{\quad}) = -4y - 3 \underline{\quad}$ Then add $\left(\dfrac{b}{2}\right)^2$ to both sides. $\left(\dfrac{b}{2}\right)^2 = \left(\dfrac{4}{2}\right)^2 = 4$ Notice that

$2(x^2 + 4x + 4) = -4y - 3 + 8$ you actually add 8 because of the multiplication of 2.

$2(x+2)(x+2) = -4y + 5$

$2(x+2)^2 = -4\left(y - \dfrac{5}{4}\right)$ Factor both sides.

$(x+2)^2 = -2\left(y - \dfrac{5}{4}\right)$ Divide by 2 to get the equation in standard form.

$h = -2 \quad 4p = -2 \quad k = \dfrac{5}{4}$ Now, find the variables for all the components.

$\quad\quad\quad p = -\dfrac{1}{2}$

Vertex: $(h, k) = \left(-2, \dfrac{5}{4}\right)$ Focus: $(h, k+p) = \left(-2, \dfrac{3}{4}\right)$ Directrix: $y = \dfrac{7}{4}$

Example 4: Find the equation of the parabola in standard form that satisfies the conditions.
Focus $(2,-4)$ and Vertex $(2,2)$

Solution: Draw a rough sketch of the given information. Notice the focus is below the vertex, therefore the parabola opens down. The standard form will be $(x-h)^2 = 4p(y-k)$. The variable p is the distance from the vertex to the focus, but we will use -6 since the graph opens down.

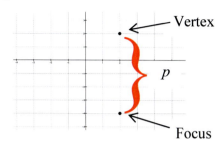

Variables:
$h = 2$
$k = 2$
$p = -6$

Equation:
$(x-2)^2 = -24(y-2)$

Example 5: Find the equation of the parabola in standard form that satisfies the conditions.
Focus $(-1,2)$ and Directrix $x = -5$

Solution: Sketch the given information. The vertex is the midpoint between the directrix and the focus. The midpoint formula requires two points. Find the point on the directrix $(-5,2)$ then use the formula. The variable p is positive since the graph opens right.

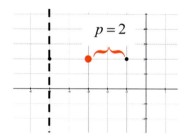

Midpoint:
$\left(\dfrac{x_1 + x_2}{2}, \dfrac{y_1 + y_2}{2}\right)$

$\left(\dfrac{-1+(-5)}{2}, \dfrac{2+2}{2}\right)$

$(-3,2)$ Vertex

Equation:
$(y-2)^2 = 8(x+3)$

Example 6: Find the equation of the parabola in standard form that satisfies the conditions.
Vertex $(-4,1)$, Point $(-2,2)$ is on the parabola, and axis of symmetry is parallel to the y-axis

Solution: Since the axis of symmetry is parallel to the y-axis, the directrix is a horizontal line. With the given information, the parabola opens up. We know the vertex, so only the value of p is unknown. Use the given point and the vertex, to solve for p.

Variables:
$h = -4$
$k = 1$
$p = ?$

$x = -2$
$y = 2$

$$(x-h)^2 = 4p(y-k)$$
$$(-2-(-4))^2 = 4p(2-1)$$
$$(2)^2 = 4p(1)$$
$$1 = p$$

The equation for the parabola is $(x+4)^2 = 4(y-1)$.

Parabolas can be used to model many applications. If you have DISH or DirectTV, the cross section of the satellite is a parabolic shape. The location of the receiver is at the focus. For headlights in a vehicle, the cross section is also a parabolic shape. The location of the bulb is at the focus. Incoming radio or light waves parallel to the axis are reflected into the focus. See exercises 21 and 22.

Example 7: The towers of the Golden Gate Bridge connecting San Francisco to Martin County are 1280 meters apart and rise 160 meters above the road. The cable between the towers has the shape of a parabola, and the cable just touches the sides of the road midway between the towers.

a. Draw a picture on the x-y axis and label the values given.

b. Find p using the given information. Determine the equation for the Golden Gate Bridge.

c. Find the height of the cable 200 meters from a tower.

Solution:

a. When placing an application for a parabola on the x-y axis, it is best to use the origin as the vertex.

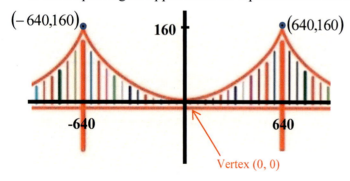

b. Since the parabola opens up, the standard form is $(x-h)^2 = 4p(y-k)$. Using the origin as the vertex, $x^2 = 4py$.

To find p, use one of the points on the parabola. For the right tower, $(640, 160)$.

$$640^2 = 4p(160)$$
$$409600 = 640p$$
$$640 = p$$

The standard form of the equation with vertex $(0,0)$ and $p = 640$ is $x^2 = 2560y$.

$$(x-0)^2 = 4(640)(y-0)$$
$$x^2 = 2560y$$

c. To find the height of the cable 200 meters from a tower, let $x = 440$.

$$440^2 = 2560y$$
$$\frac{193600}{2560} = y$$

So, the height of the cable will be 75.625 meters.

FINAL EXAM REVIEW

KEEP IT FRESH

1. $f(x) = \dfrac{x^2 + x - 2}{x - 3}$

   a. What is the domain of the function?

   b. What is the vertical asymptote?

   c. What is the slant asymptote?

   d. What are the x-intercepts?

   e. What is the y-intercept?

   g. Graph the rational function.

2. Find the coordinates (x, y) of the hole in the rational function.

   $f(x) = \dfrac{x^3 - x^2}{x - 1}$

Lesson 18　　　　　　　　　　　　　　　　　　　　　　Practice Exercises

In Exercises 1 – 4, classify the equation as that of a circle, ellipse, hyperbola, or parabola.

1. $4x^2 - 9y^2 - 24x + 72y - 144 = 0$
2. $16x^2 + 5y^2 - 3x + 4y - 538 = 0$
3. $(x-1)^2 + (y+1)^2 = 100$
4. $2x^2 - 8x + y + 9 = 0$

In Exercises 5 – 10, find the equation of the parabola in standard form that satisfies the conditions.

5. Focus $(0,4)$ and Directrix $y = -4$
6. Focus $(1,1)$ and Directrix $y = 6$
7. Focus $(-3,2)$ and Directrix $x = 0$
8. Focus $(-4,0)$ and Directrix $x = 4$
9. Vertex $(2,-5)$, axis of symmetry is $x = 2$, and $(4,-3)$ is a point on the parabola
10. Vertex $(-1,4)$, directrix is parallel to the y-axis, and $(3,0)$ is a point on the parabola

In Exercises 11 – 14, identify the vertex, focus, directrix, and the x & y – intercepts. Then graph.

11. $y^2 = 20x$
12. $(y-3)^2 = -8(x+1)$
13. $x^2 = -4y$
14. $(x+2)^2 = 16(y+3)$

In Exercises 15 – 20, write the equation of the parabola in standard form. Identify the vertex, focus, directrix, and endpoints of the latus rectum. Then sketch the graph of the parabola.

15. $y^2 + 8y - 4x + 8 = 0$
16. $x^2 + 6x - 12y + 33 = 0$
17. $2x^2 + 8x + y + 9 = 0$
18. $y^2 - 4y + 12x - 20 = 0$
19. $x^2 + 4x - y = 0$
20. $3y^2 - 6y - 24x - 21 = 0$

21. A satellite dish has the shape of a parabola. The signals that it receives are reflected to a receiver that is located at the focus of the parabola. If the dish is 8 feet across at its opening and 1.25 feet deep at its center, determine the location of its receiver.

22. A light source is to be placed on the axis of symmetry of the parabolic reflector. How far to the right of the vertex point should the light source be located if the designer wishes the reflected light rays to form a beam of parallel rays?

Solutions for Practice Exercises Lesson 18

1. Hyperbola 2. Ellipse 3. Circle 4. Parabola

5. $x^2 = 16y$ 6. $(x-1)^2 = -10(y-3.5)$ 7. $(y-2)^2 = -6(x+1.5)$ 8. $y^2 = -16x$

9. $(x-2)^2 = 2(y+5)$ 10. $(y-4)^2 = 4(x+1)$

11. Vertex $(0,0)$; Focus $(5,0)$

Directrix $x = -5$; x-int $(0,0)$; y-int $(0,0)$

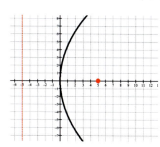

12. Vertex $(-1,3)$; Focus $(-3,3)$

Directrix $x = 1$; x-int $\left(-\dfrac{17}{8}, 0\right)$; no y-int

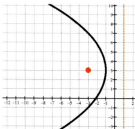

13. Vertex $(0,0)$; Focus $(0,-1)$

Directrix $y = 1$; x-int $(0,0)$; y-int $(0,0)$

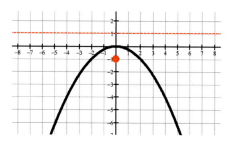

14. Vertex $(-2,-3)$; Focus $(-2,1)$

Directrix $y = -7$; x-int $(-2 \pm 4\sqrt{3}, 0)$; y-int $\left(0, -\dfrac{11}{4}\right)$

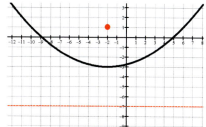

15. $(y+2)^2 = 4(x+2)$; Vertex $(-2,-4)$

Focus $(-1,-4)$; Directrix $x = -3$

Focal Chord Endpts $(-1,-2)$ & $(-1,-6)$

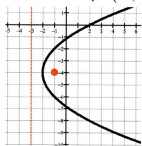

16. $(x+3)^2 = 12(y-2)$; Vertex $(-3,2)$

Focus $(-3,5)$; Directrix $y = -1$

Focal Chord Endpts $(-9,5)$ & $(3,5)$

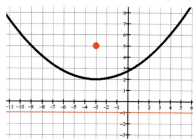

17. $(x+2)^2 = -\frac{1}{2}(y+1)$; Vertex $(-2,-1)$

Focus $\left(-2,-\frac{9}{8}\right)$; Directrix $y = -\frac{7}{8}$

Focal Chord Endpts $\left(-\frac{7}{4},-\frac{9}{8}\right) \& \left(-\frac{9}{4},-\frac{9}{8}\right)$

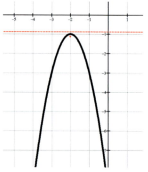

18. $(y-2)^2 = -12(x-2)$; Vertex $(2,2)$

Focus $(-1,2)$; Directrix $x = 5$

Focal Chord Endpts $(-1,8) \& (-1,-4)$

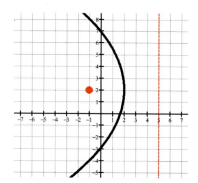

19. $(x+2)^2 = y+4$; Vertex $(-2,-4)$

Focus $\left(-2,-\frac{15}{4}\right)$; Directrix $y = -\frac{17}{4}$

Focal Chord Endpts $\left(-\frac{3}{2},-\frac{15}{4}\right) \& \left(-\frac{5}{2},-\frac{15}{4}\right)$

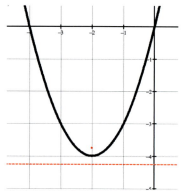

20. $(y-1)^2 = 8(x+1)$; Vertex $(-1,1)$

Focus $(1,1)$; Directrix $x = -3$

Focal Chord Endpts $(1,5) \& (1,-3)$

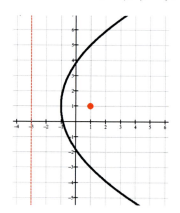

21. The receiver should be 3.2 ft from the vertex.

22. The light source should be 0.375 inches to the right of the vertex.

Cumulative Review 4

Lessons 15 – 18

1. Factor completely.

   a. $x(1+x)^{-1/2} + 2(1+x)^{1/2}$

   b. $(x+1)^{1/3} + x\left(\dfrac{1}{3}\right)(x+1)^{-2/3}$

   c. $\dfrac{(x+4)^{1/2} - 2x(x+4)^{-1/2}}{x+4}$

   d. $\dfrac{10x(x^2+4)^{-5}(x-6)^{1/4} + 4(x^2+4)^{-4}(x-6)^{-3/4}}{(x^2+4)^3}$

2. Find the center, vertices, foci, length of major axis, and eccentricity. Then graph the conic section. Be sure to be specific with plotting points on the graph.

   a. $\dfrac{(x+2)^2}{9} + \dfrac{(y-1)^2}{25} = 1$

   b. $x^2 + 9y^2 + 6x - 36y + 36 = 0$

3. Find the equation in standard from of the ellipse with vertices (–6,4) and (2,4), foci (–5,4) and (1,4).

4. Find the equation of the ellipse in standard form with vertices (–2,4) and (–2,–2), and passing through (0,1).

5. An arch of a bridge has the shape of the top half of an ellipse. The arch is 40 ft wide and 12 ft high at the center. Find the equation of the complete ellipse. Find the height of the arch 10 ft from the center of the bottom.

6. Find the center, vertices, foci, asymptotes, and eccentricity. Then graph the conic section. Be sure to be specific with plotting points on the graph.

   a. $\dfrac{(y-3)^2}{16} - \dfrac{(x-2)^2}{9} = 1$

   b. $4x^2 - y^2 + 32x + 6y + 39 = 0$

7. Find the equation of the hyperbola in standard form that has vertices of (9,1) and (–3,1), and the slope of an asymptote is $\dfrac{1}{2}$.

8. Find the equation of the hyperbola in standard form that has vertices of (0,–4) and (0,4) and foci of (0, –5) and (0,5).

9. Find the vertex, focus, and directrix. Then graph the conic section. Be sure to be specific with plotting points on the graph.

   a. $2x - y^2 - 6y + 1 = 0$

   b. $2y + x^2 + 8x + 8 = 0$

10. Find the equation in standard form of the parabola with focus (–3,0) and directrix $y = 2$.

11. Find the equation in standard form of the parabola with focus (5,–1) and directrix $x = -5$.

12. A car headlight mirror has a parabolic cross section with diameter 6 in and depth 1 in. How far from the vertex should the bulb be positioned if it is to be placed at the focus?

Solutions to Review

1a. $\dfrac{3x+2}{(1+x)^{1/2}}$ 1b. $\dfrac{4x+3}{3(x+1)^{2/3}}$ 1c. $\dfrac{-x+4}{(x+4)^{3/2}}$ 1d. $\dfrac{2(7x-2)(x-4)}{(x^2+4)^8(x-6)^{3/4}}$

2a.
$(-2,1),\ (-2,-4)(-2,6),$
$(-2,-3)(-2,5),\ \text{Vertical}\ 2b=10$

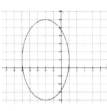
$e=\dfrac{4}{5}$

2b.
$(-3,2),\ (0,2)(-6,2),$
$(-3\pm 2\sqrt{2},2),\ \text{Horizontal}\ 2a=6$

$e=\dfrac{2\sqrt{2}}{3}$

3. $\dfrac{(x+2)^2}{16}+\dfrac{(y-4)^2}{7}=1$ 4. $\dfrac{(x+2)^2}{4}+\dfrac{(y-1)^2}{9}=1$ 5. $\dfrac{x^2}{400}+\dfrac{y^2}{144}=1$; 10.39 ft

6a.
$(2,3),\ (2,-1)(2,7),\ (2,-2)(2,8),$
$y-3=\pm\dfrac{4}{3}(x-2),\ e=\dfrac{5}{4}$

6b.
$(-4,3),\ (-6,3)(-2,3),\ (-4\pm 2\sqrt{5},3),$
$y-3=\pm 2(x+4),\ e=\sqrt{5}$

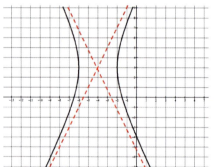

7. $\dfrac{(x-3)^2}{36}-\dfrac{(y-1)^2}{9}=1$ 8. $\dfrac{y^2}{16}-\dfrac{x^2}{9}=1$

9a. $(-5,-3),\ \left(-\dfrac{9}{2},-3\right),\ x=-\dfrac{11}{2}$ 9b. $(-4,4),\ \left(-4,\dfrac{7}{2}\right),\ y=\dfrac{9}{2}$

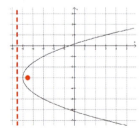 

10. $(x+3)^2=-4(y-1)$ 11. $(y+1)^2=20x$ 12. 9/4 in.

Lesson 19   Sequences and Series

The following set of numbers may look like a random list but a closer glance may reveal a pattern. Can you determine the pattern?

1, 3, 6, 10, 15, 21, 28, 36, 45

In this lesson, we will study lists of numbers to determine if a pattern exists. The pattern can be described as a **sequence**, a function whose domain is the set of consecutive natural numbers (positive integers beginning with 1).

Compare function notation with sequence notation. The variable x is replaced with n. The name of a sequence uses the letter a instead of f. Your calculator has a sequence mode. Press MODE on your calculator, move your cursor to SEQ, and press ENTER.

Function Notation   **Sequence Notation**

$f(x) = 2x - 5$ $a_n = 2n - 5$

$D: (-\infty, \infty)$ $D: \{1, 2, 3, 4, ...\}$

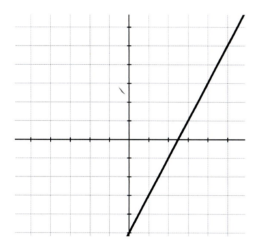
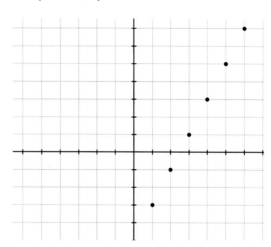

$a_n = 2n - 5$ is called the **general term**, which is used to find all the terms of the sequence by evaluating values for n. To find the value of the first term a_1, evaluate the general term for $n = 1$.

$a_1 = 2(1) - 5 = -3$ The point $(1, -3)$ is plotted on the graph of the sequence.

You can find the 2nd and 3rd terms by evaluating for $n = 2$ and $n = 3$.

$a_2 = 2(2) - 5 = -1$ $a_3 = 2(3) - 5 = 1$

After finding the first 3 terms of a sequence, a pattern has been established. Write the terms of the sequence in a list, and you can see the numbers differ by 2. $-3, -1, 1, ...$

Since the points are not connected, a sequence is discontinuous. There are several types of sequences that we will study in this lesson. First, an **infinite sequence** is a sequence whose domain is $D: \{1, 2, 3, 4, ...\}$ and, a **finite sequence** is a sequence whose domain is $D: \{1, 2, 3, 4, ..., n\}$.

Infinite Sequence $-3, -1, 1, ...$ **Finite sequence** $-3, -1, 1, ..., 25$

Can you determine how many terms are in the finite sequence, $-3, -1, 1, ..., 25$? In other words, what term number is 25? To answer this question, you need to find the value of n that makes $a_n = 25$. Set the general term equal to 25 and solve for n.

$2n - 5 = 25$ Add 5 and divide by 2.
$n = 15$

The 15th term is 25, so $a_{15} = 25$. Therefore, the finite sequence has 15 terms.

Example 1: **Find the first 4 terms of the sequence. Write a finite sequence for the first 12 terms.**

$a_n = (-1)^n 2n$

Solution: Evaluate the sequence for $n = 1, 2, 3, 4,$ and 12.

$a_1 = (-1)^1 2(1) = -2$

$a_2 = (-1)^2 2(2) = 4$

$a_3 = (-1)^3 2(3) = -6$

$a_4 = (-1)^4 2(4) = 8$

$a_{12} = (-1)^{12} 2(12) = 24$ Finite sequence with 12 terms would be listed as $-2, 4, -6, 8, ..., 24$.

Notice the terms in Example 1 have opposite signs. In fact, the sign of the number depends on the power of -1. This is an example of an **alternating sequence,** a sequence whose terms alternate signs between positive and negative. For Example 1, the odd terms are negative and the even terms are positive. Can you think of an alternating sequence whose 1st term is positive and 2nd term is negative?

$(-1)^{n+1}$ or $(-1)^{n-1}$ Remember: the sign of the number depends on the power of -1.

Now that you can find the terms of a sequence, let's see if we can find the general term. The general term is a function using the variable n instead of x. Remember all the functions you have studied: linear, quadratic, rational, exponential, etc. Any of these functions can be used to write the general term of the sequence.

Example 2: Find the general term for the following infinite sequences. Then find the 20th term.

a. $0, 1, 4, 9, \ldots$

b. $\dfrac{1}{2}, -\dfrac{1}{6}, \dfrac{1}{12}, -\dfrac{1}{20}, \ldots$

Solution:

a. Find a pattern by determining the next term, a_5.

$a_1 = 0 \qquad a_2 = 1 \qquad a_3 = 4 \qquad a_4 = 9 \qquad a_5 = \underline{\;?\;}$

Since the numbers in the sequence are perfect squares, the 5th term in the list would be $a_5 = 16$. Therefore, we will need a quadratic function. If you evaluate n^2 for 1, the result is not 0 as the 1st term requires. We need a horizontal transformation to shift the values. By trial and error, the formula $a_n = (n-1)^2$ is the general term that produces the sequence.

The 20th term is $a_{20} = (20-1)^2 = 361$.

---

b. The general term will definitely include a fraction with numerator 1. Therefore, a rational function is used to generate the terms in the sequence. Also, notice the terms alternate. We will need to use a power of -1. Since the first term is positive, we should use $(-1)^{n+1}$ or $(-1)^{n-1}$. Now, focus on the numbers in the denominator. What is the pattern?

$a_1 = \dfrac{1}{\boxed{2}} \qquad a_2 = -\dfrac{1}{\boxed{6}} \qquad a_3 = \dfrac{1}{\boxed{12}} \qquad a_4 = -\dfrac{1}{\boxed{20}}$

Is there a consistent number that is multiplied?   2 6 12 20

No. $2 \cdot 3 = 6 \quad 6 \cdot 2 = 12 \quad 12 \cdot \dfrac{5}{3} = 20$

Is there a consistent number that is added?   2 6 12 20

No, but there is a pattern. $+4 \quad +6 \quad +8$

We can guess the next denominator is 30 (20 + 10). Therefore, the 5th term is $a_5 = \dfrac{1}{30}$.

To find a formula for the general term, you need to find a relationship between the term number and the number in the sequence. Ask the question: How does $n=1$ produce the number 2? How does $n=2$ produce the number 6? Start with basic operations like addition and multiplication.

For $n=1$, $1+1=2$ and $1\cdot 2=2$.
        $+1^2$ $1(1+1)$

For $n=2$, $2+4=6$ and $2\cdot 3=6$.
        $+2^2$ $2(2+1)$

For $n=3$, $3+9=12$ and $3\cdot 4=12$.
        $+3^2$ $3(3+1)$

For $n=4$, $4+16=20$ and $4\cdot 5=20$.
        $+4^2$ $4(4+1)$

For $n=5$, $5+25=30$ and $5\cdot 6=30$.
        $+5^2$ $5(5+1)$

Both of these operations indicate a pattern to use for the general term. For the general variable n, $n+n^2$ and $n\cdot(n+1)$ generate the number in the denominator of each term. By trial and error, $a_n = (-1)^{n+1}\dfrac{1}{n+n^2}$ or $a_n = (-1)^{n+1}\dfrac{1}{n(n+1)}$ would work as the general term that produces the sequence. The 20th term is $a_{20} = -\dfrac{1}{420}$.

When finding the general term, realize that the answer is not unique. For Example 2b, the general term $a_n = \dfrac{(-1)^{n-1}}{n^2+n}$ could also be used to generate all the terms of the sequence. If your answer doesn't match the answer in the key, don't panic. Your answer could be correct, as long as the formula produces every term in the sequence.

Recursion Formula or Recursively Defined Sequence

A sequence that lists the 1st term (or the first few terms) and describes how to determine the remaining terms from the given term is called a **recursively defined sequence**.

Example 3: Find the first five terms of the sequence.

a. $a_n = \begin{cases} a_1 = -3 \\ a_n = n a_{n-1} \end{cases}$ **b.** $a_n = \begin{cases} a_1 = 1 \text{ and } a_2 = 1 \\ a_n = a_{n-2} + a_{n-1} \end{cases}$

Solution:

a. Since the 1st term is given, we need to start with the 2nd term. Using the formula given, find the value when $n=2$.

$a_n = n\,a_{n-1}$

$a_2 = 2\,a_{2-1}$ $a_1 = -3$

$ = 2\,a_1$

$ = 2(-3)$

$a_2 = -6$

Notice: a recursion formula depends on the preceding terms not just the variable n itself. If you were asked to find the 20th term, you would need the 19th term. If you need the 19th term, then you also need the 18th term. And so on! This is one disadvantage to using a recursively defined sequence. You cannot find any random term number without first knowing all the preceding numbers.

To find the 3rd term, find the value of the formula for $n = 3$.

$a_n = n\,a_{n-1}$

$a_3 = 3\,a_{3-1}$ $a_2 = -6$

$ = 3\,a_2$

$ = 3(-6)$

$a_3 = -18$

To find the 4th term, find the value of the formula for $n = 4$.

$a_n = n\,a_{n-1}$

$a_4 = 4\,a_{4-1}$ $a_3 = -18$

$ = 4\,a_3$

$ = 4(-18)$

$a_4 = -72$

The 5th term is $a_5 = -360$. The first 5 terms of the recursively defined sequence are $-3, -6, -18, -72, -360$.

---

b. For this recursively defined sequence, the first two terms are given.

To find the 3rd term, find the value of the formula for $n = 3$.

$a_n = a_{n-2} + a_{n-1}$

$a_3 = a_{3-2} + a_{3-1}$ $a_1 = 1$ and $a_2 = 1$

$ = a_1 + a_2$

$ = 1 + 1$

$a_3 = 2$

To find the 4th term, find the value of the formula for $n = 4$.

$a_n = a_{n-2} + a_{n-1}$

$a_4 = a_{4-2} + a_{4-1}$ $a_2 = 1$ and $a_3 = 2$

$ = a_2 + a_3$

$ = 1 + 2$

$a_4 = 3$

To find the 5th term, find the value of the formula for $n=5$.

$a_n = a_{n-2} + a_{n-1}$
$a_5 = a_{5-2} + a_{5-1}$
$\quad\;\; = a_3 + a_4 \qquad a_3 = 2$ and $a_4 = 3$
$\quad\;\; = 2 + 3$
$a_5 = 5$

The first 5 terms of the recursively defined sequence are $1, 1, 2, 3, 5$.

One of the most famous recursively defined sequence is associated with the work of Leonardo of Pisa, better known as Fibonacci. The sequence in Example 3b is known as the Fibonacci Sequence, each term is the sum of the previous two terms, beginning with 1 and 1.

1, 1, 2, 3, 5, 8, 13, 21, 34, 55,...

The sum of the terms in a sequence is called a **series**. A **finite series** (also known as the partial sum) is the sum of the first n terms $S_n = a_1 + a_2 + a_3 + a_4 + ... + a_n$. An **infinite series** is the sum $S_\infty = a_1 + a_2 + a_3 + ... + a_n + ...$ We will study the infinite series in Lesson 21. For now, let's focus on the partial sums.

Example 4: For the following sequence, find the partial sums S_4, S_6, and S_k.

$7, 10, 13, 16, ... 3n + 4$

Solution:

$S_4 = a_1 + a_2 + a_3 + a_4$
$\quad\;\; = 7 + 10 + 13 + 16$
$\quad\;\; = 46$

$S_6 = a_1 + a_2 + a_3 + a_4 + a_5 + a_6$
$\quad\;\; = 7 + 10 + 13 + 16 + 19 + 22$
$\quad\;\; = 87$

$S_k = a_1 + a_2 + a_3 + ... + a_k$
$\quad\;\; = 7 + 10 + 13 + ...(3k + 4)$

general term $a_n = 3n + 4$

$a_5 = 3(5) + 4 = 19$
$a_6 = 3(6) + 4 = 22$
$a_k = 3(k) + 4 = 3k + 4$

Sigma Notation or Summation Notation

If the general term of the sequence is known, the Greek letter sigma Σ can be used to write a series.

In Example 4, the general term is $a_n = 3n+4$. Each partial sum can be written in sigma notation.

$$S_4 = \sum_{i=1}^{4} 3i+4$$

upper bound — 4
ith term (general term)
i is the index of summation
lower bound — $i=1$

Other common variables are j, k, and n.

This sigma notation indicates the sum of the terms $a_1 + a_2 + a_3 + a_4$. The terms are generated as the index of summation cycles from 1 to 4 into the ith term.

Example 5: Compute each sum.

a. $\sum_{i=1}^{6} i^2$

b. $\sum_{k=1}^{5} \frac{(-1)^k}{k}$

Solution:

a. Find the sum of the first 6 perfect squares. $\sum_{i=1}^{6} i^2 = 1^2 + 2^2 + 3^2 + 4^2 + 5^2 + 6^2$

$$= 1 + 4 + 9 + 16 + 25 + 36$$
$$= 91$$

b. This partial sum is called an alternating series. Notice the terms alternate between positive and negative signs.

$$\sum_{k=1}^{5} \frac{(-1)^k}{k} = \frac{(-1)^1}{1} + \frac{(-1)^2}{2} + \frac{(-1)^3}{3} + \frac{(-1)^4}{4} + \frac{(-1)^5}{5}$$

$$= -1 + \frac{1}{2} - \frac{1}{3} + \frac{1}{4} - \frac{1}{5}$$

$$= -\frac{47}{60}$$

You can find the sum of a sequence using a TI-84. First, make sure you are in sequence mode. Then press 2ND and STAT. Under the MATH submenu, press 5. Now press 2ND and STAT again. Under the OPS submenu, press 5.

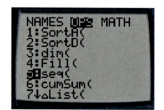

For Example 5a, enter the expression n^2. The variable is n. For this example, we start at 1 and end at 6 with a step of 1.

Variable Key: For sequence mode, the variable n will be used.

Example 6: Write the indicated sum using sigma notation.

a. $1 + \dfrac{1}{2} + \dfrac{1}{4} + \dfrac{1}{8} + \dfrac{1}{16} + \dfrac{1}{32}$

b. $1 + 2 + 3 + 4 + 5 + \ldots + 10$

Solution:

a. Using a lower bound of 1, find a pattern for the general term. Each term can be written as a fraction with numerator 1. The denominator changes by a multiple of 2. 2^n would result in the numbers 2, 4, 8, ... The first term in the sequence would need to include $2^0 = 1$. Therefore, the numbers in the denominator result from the expression 2^{n-1}. Now, put the numerator and denominator together to form the general term. Use the variable i for the index of summation.

$$\sum_{i=1}^{6} \dfrac{1}{2^{i-1}}$$

b. The numbers in the series are counting numbers. The general term would be $a_n = n$.

$a_1 = 1$, $a_2 = 2$, and so on. Use the variable i for the index of summation.

$$\sum_{i=1}^{10} i$$

These answers are not unique. The lower bound does not have to be 1. Any integer less than or equal to the upper bound would work. Take a look at Example 6b. Another correct answer would be $\sum_{k=0}^{9} k+1$. If the lower bound is 0, the general term and the upper bound must indicate the shift. Would the sigma notation $\sum_{k=2}^{11} k-1$ result in the same sum? Yes!

$$\sum_{k=2}^{11} k-1 = a_2 + a_3 + a_4 + \ldots + a_{11}$$
$$= (2-1) + (3-1) + (4-1) + \ldots + (11-1)$$
$$= 1 + 2 + 3 + \ldots + 10$$

Properties of Summation

1. $\sum_{i=1}^{n} c = cn$ — Adding the constant c "n" times will result in cn.

2. $\sum_{i=1}^{n} ca_i = c \sum_{i=1}^{n} a_i$ — A constant can be factored out of a sum.

3. $\sum_{i=1}^{n} (a_i \pm b_i) = \sum_{i=1}^{n} a_i \pm \sum_{i=1}^{n} b_i$ — The sigma can be distributed to two sequences.

4. $\sum_{i=1}^{m} a_i + \sum_{i=m+1}^{n} a_i = \sum_{i=1}^{n} a_i$ for $1 \leq m < n$

The summation can be condensed if the bounds are consecutive.

Examples of Properties

1. $\sum_{i=1}^{10} 3 = 3(10) = 30$

2. $\sum_{i=1}^{7} 4i = 4 \sum_{i=1}^{7} i = 4(1+2+3+\ldots+7) = 112$

3. $\sum_{i=1}^{3} (i^4 - i) = \sum_{i=1}^{3} i^4 - \sum_{i=1}^{3} i = (1^4 + 2^4 + 3^4) - (1+2+3) = 92$

4. $\sum_{i=1}^{5} 2^i + \sum_{i=6}^{10} 2^i = \sum_{i=1}^{10} 2^i = 2^1 + 2^2 + 2^3 + \ldots + 2^{10} = 2046$

Factorial

Let n be a natural number (positive integer). Then n **factorial** is defined as $n! = 1 \cdot 2 \cdot 3 \cdot 4 \cdots (n-1) \cdot n$. As a special case, zero factorial is defined as $0! = 1$.

$1! = 1$

$2! = 1(2) = 2$

$3! = 1(2)(3) = 6$

$4! = 1(2)(3)(4) = 24$

$5! = 1(2)(3)(4)(5) = 120$

Factorials follow the same order of operations as exponents. PEMDAS ← factorial

Note the difference between $2n!$ and $(2n)!$

For $2n!$, the factorial is done first. For $(2n)!$, multiplication by 2 is done first.

$n = 3$: $\quad 2n! = 2 \cdot 3! = 2 \cdot 6 = 12 \quad$ and $\quad (2n)! = (2 \cdot 3)! = 6! = 720$.

The calculator has a factorial operation. For the TI-84, press MATH and move the cursor to the PRB submenu. Press 4.

Example 7: Simplify the factorial expressions.

a. $\dfrac{12!}{10!}$ **b.** $\dfrac{6!}{3!3!}$

Solutions:

a. $\dfrac{12!}{10!} = \dfrac{\cancel{1(2)(3)\dots(10)}(11)(12)}{\cancel{1(2)(3)\dots(10)}} = 11(12) = 132$

b. $\dfrac{6!}{3!3!} = \dfrac{\cancel{1(2)(3)}(4)(5)\cancel{(6)}}{\cancel{1(2)(3)} \cdot 1\cancel{(2)(3)}} = 4(5) = 20$

Example 8: Find the first 3 terms of the sequence.

a. $a_n = \dfrac{2n!}{(n+1)!}$ **b.** $\dfrac{(n-1)!}{(n+2)!}$

Solutions:

a. You could start by evaluating the general term for values of n. But, you will find the evaluation more efficient if you simplify the expression first.

$$a_n = \frac{2n!}{(n+1)!} = \frac{2 \cdot \cancel{1(2)(3)\dots n}}{\cancel{1(2)(3)\dots n}(n+1)} = \frac{2}{n+1}$$

Using the reduced formula, find the value when $n = 1$. $a_1 = \dfrac{2}{1+1} = 1$

For $n = 2$, $a_2 = \dfrac{2}{2+1} = \dfrac{2}{3}$. For $n = 3$, $a_3 = \dfrac{2}{3+1} = \dfrac{2}{4} = \dfrac{1}{2}$.

The first 3 terms of the sequence are $1, \dfrac{2}{3}, \dfrac{1}{2}$.

b. Simplify the expression first.

$$\frac{(n-1)!}{(n+2)!} = \frac{\cancel{1(2)(3)\dots(n-1)}}{\cancel{1(2)(3)\dots(n-1)}n(n+1)(n+2)} = \frac{1}{n(n+1)(n+2)}$$

Now evaluate for values of n.

$$a_1 = \frac{1}{n(n+1)(n+2)} = \frac{1}{1(1+1)(1+2)} = \frac{1}{6}$$

$$a_2 = \frac{1}{n(n+1)(n+2)} = \frac{1}{2(2+1)(2+2)} = \frac{1}{24}$$

$$a_3 = \frac{1}{n(n+1)(n+2)} = \frac{1}{3(3+1)(3+2)} = \frac{1}{60}$$

The first 3 terms of the sequence is $\frac{1}{6}, \frac{1}{24}, \frac{1}{60}$.

FINAL EXAM REVIEW

KEEP IT FRESH

1. Write the augmented matrix for the system of equations. Use Gaussian or Gauss-Jordan elimination to find the solution.

$$\begin{cases} -x+3y+2z=-10 \\ 3x-2y-2z=7 \\ -2x+y-z=-10 \end{cases}$$

2. Decompose the rational expression into partial fractions.

a. $\dfrac{3x^2+17x+12}{(x-1)(x+3)^2}$

b. $\dfrac{x^2-x-7}{(x+1)(x^2+4)}$

Lesson 19   Practice Exercises

In Exercises 1 – 6, find the indicated term of the given sequence.

1. $a_n = 4n - 2$; a_3

2. $a_n = \dfrac{1}{2}n^2 + 1$; a_4

3. $a_n = \dfrac{n}{n-1}$; a_{10}

4. $a_n = 2^{n+1} + 7$; a_2

5. $a_n = (-1)^n n$; a_9

6. $a_n = \dfrac{(-1)^{n+1}}{n^2+1}$; a_7

In Exercises 7 – 15, find the first five terms of the sequence.

7. $a_n = \left(\dfrac{2}{3}\right)^n$

8. $a_n = (-n)^3 - 1$

9. $a_n = \dfrac{3}{4}n + 6$

10. $a_n = \dfrac{n^2}{n+1}$

11. $a_n = \dfrac{(-1)^n}{2n-1}$

12. $a_n = 3^{2n-1} - 1$

13. $a_n = \begin{cases} a_1 = -2 \\ a_n = 3a_{n-1} + 5 \end{cases}$

14. $a_n = \begin{cases} a_1 = 1 \\ a_n = (a_{n-1})^2 - 4 \end{cases}$

15. $a_n = \begin{cases} a_1 = 3, a_2 = 7 \\ a_n = -2a_{n-1} + 2a_{n-2} \end{cases}$

In Exercises 16 – 21, find the general term of each sequence. Do not use a recursion formula. Then find the 20th term.

16. $0, 1, 8, 27, \ldots$

17. $0, 3, 8, 15, \ldots$

18. $-\dfrac{1}{2}, \dfrac{2}{3}, -\dfrac{3}{4}, \dfrac{4}{5}, \ldots$

19. $3, -9, 27, -81, \ldots$

20. $5, 10, 15, 20, \ldots$

21. $1, e, e^2, e^3, \ldots$

In Exercises 22 – 27, find the indicated partial sum for each sequence.

22. $a_n = 3n - 11$; S_6

23. $a_n = \dfrac{n-1}{n+1}$; S_4

24. $a_n = (n-1)^2 + 2$; S_5

25. $a_n = (-2)^{n-2}$; S_5

26. $a_n = \left(\dfrac{1}{4}\right)^n - 5$; S_3

27. $a_n = n + 8$; S_{10}

In Exercises 28 – 33, compute each sum.

28. $\displaystyle\sum_{k=1}^{12} 6k - 1$

29. $\displaystyle\sum_{k=4}^{11} 5k + 6$

30. $\displaystyle\sum_{k=1}^{4} \dfrac{(-1)^{k+1}}{k(k+1)}$

31. $\displaystyle\sum_{i=3}^{7} 2i^3$

32. $\displaystyle\sum_{i=2}^{10} (-1)^i + 1$

33. $\displaystyle\sum_{i=1}^{5} \dfrac{2i}{i^2+4}$

In Exercises 34 – 35, expand the sigma notation using the properties of summation. Then evaluate the series.

34. $\sum_{k=1}^{10} 2k^2 - k$

35. $\sum_{i=1}^{7} 5i + 9$

In Exercises 36 – 37, condense the sigma notation using the 4th property of summation. Then evaluate the series.

36. $\sum_{k=1}^{4} 3k^2 + \sum_{k=5}^{10} 3k^2$

37. $\sum_{k=1}^{3} \frac{1}{k} + \sum_{k=4}^{8} \frac{1}{k}$

In Exercises 38 – 45, write the indicated sum using sigma notation.

38. $a_n = -6n$; S_{12}

39. $a_n = (-1)^n 2^{n+1} - n$; S_6

40. $a_n = \frac{1}{n^3}$; S_7

41. $1 + 3 + 5 + \ldots + 21$

42. $\frac{1}{2} - \frac{1}{4} + \frac{1}{6} - \frac{1}{8} + \frac{1}{10} - \frac{1}{12} + \frac{1}{14}$

43. $2 + 9 + 28 + 65 + 126 + 217 + 344 + 513 + 730 + 1001$

44. $1 + 0.1 + 0.01 + 0.001 + \ldots$

45. $-2 + 4 - 8 + 16 - \ldots$

In Exercises 46 – 49, simplify the factorial expressions.

46. $\frac{5!}{8!}$

47. $\frac{(5+1)!}{(9-2)!}$

48. $\frac{2 \cdot 4!}{7!}$

49. $\frac{9!}{(2 \cdot 3)!}$

In Exercises 50 – 53, find the first five terms of the sequence.

50. $a_n = \frac{2n!}{(n+1)!}$

51. $a_n = \frac{(2n)!}{(2n+1)!}$

52. $a_n = \frac{(n-1)!}{n!}$

53. $a_n = \frac{(n+2)!}{n!}$

In Exercises 54 – 55, write the indicated sum using sigma notation.

54. $1 + 1 + 2 + 6 + \ldots + 5040$

55. $1 + \frac{1}{2} + \frac{1}{6} + \frac{1}{24} + \frac{1}{120} + \frac{1}{720}$

Solutions to Practice Exercises Lesson 19

1. $a_3 = 10$ 2. $a_4 = 9$ 3. $a_{10} = \dfrac{10}{9}$

4. $a_2 = 15$ 5. $a_9 = -9$ 6. $a_7 = \dfrac{1}{50}$

7.
$a_1 = \dfrac{2}{3}$
$a_2 = \dfrac{4}{9}$
$a_3 = \dfrac{8}{27}$
$a_4 = \dfrac{16}{81}$
$a_5 = \dfrac{32}{243}$

8.
$a_1 = -2$
$a_2 = -9$
$a_3 = -28$
$a_4 = -65$
$a_5 = -126$

9.
$a_1 = \dfrac{27}{4}$
$a_2 = \dfrac{15}{2}$
$a_3 = \dfrac{33}{4}$
$a_4 = 9$
$a_5 = \dfrac{39}{4}$

10.
$a_1 = \dfrac{1}{2}$
$a_2 = \dfrac{4}{3}$
$a_3 = \dfrac{9}{4}$
$a_4 = \dfrac{16}{5}$
$a_5 = \dfrac{25}{6}$

11.
$a_1 = -1$
$a_2 = \dfrac{1}{3}$
$a_3 = -\dfrac{1}{5}$
$a_4 = \dfrac{1}{7}$
$a_5 = -\dfrac{1}{9}$

12.
$a_1 = 2$
$a_2 = 26$
$a_3 = 242$
$a_4 = 2186$
$a_5 = 19682$

13.
$a_1 = -2$
$a_2 = -1$
$a_3 = 2$
$a_4 = 11$
$a_5 = 38$

14.
$a_1 = 1$
$a_2 = -3$
$a_3 = 5$
$a_4 = 21$
$a_5 = 437$

15.
$a_1 = 3$
$a_2 = 7$
$a_3 = -8$
$a_4 = 30$
$a_5 = -76$

16. $a_n = (n-1)^3$
$a_{20} = 6859$

17. $a_n = n^2 - 1$
$a_{20} = 399$

18. $a_n = \dfrac{(-1)^n n}{n+1}$
$a_{20} = \dfrac{20}{21}$

19. $a_n = (-1)^{n+1} \cdot 3^n$
$a_{20} = -3^{20}$

20. $a_n = 5n$
$a_{20} = 100$

21. $a_n = e^{n-1}$
$a_{20} = e^{19}$

22. -3 23. $\dfrac{43}{30}$ 24. 40 25. $-\dfrac{11}{2}$

26. $-\dfrac{939}{64}$ 27. 135 28. 456 29. 348

30. $\dfrac{11}{30}$ 31. 1550   32. 10   33. $\dfrac{7941}{3770}$

34. $2\sum_{k=1}^{10} k^2 - \sum_{k=1}^{10} k = 770 - 55 = 715$ 35. $5\sum_{i=1}^{7} i + \sum_{i=1}^{7} 9 = 140 + 63 = 203$

36. $\sum_{k=1}^{10} 3k^2 = 1155$ 37. $\sum_{k=1}^{8} \dfrac{1}{k} = \dfrac{761}{280}$ 38. $\sum_{k=1}^{12} -6k$ 39. $\sum_{k=1}^{6} (-1)^k 2^{k+1} - k$

40. $\sum_{k=1}^{7} \dfrac{1}{k^3}$ 41. $\sum_{k=1}^{11} 2k-1$ 42. $\sum_{k=1}^{7} \dfrac{(-1)^{k+1}}{2k}$ 43. $\sum_{k=1}^{10} k^3 + 1$

44. $\sum_{k=1}^{\infty} 10^{1-k}$ 45. $\sum_{k=1}^{\infty} (-2)^k$ 46. $\dfrac{1}{336}$ 47. $\dfrac{1}{7}$

48. $\dfrac{1}{105}$ 49. 504

50. $a_n = \dfrac{2}{n+1}$ 51. $a_n = \dfrac{1}{2n+1}$ 52. $a_n = \dfrac{1}{n}$ 53. $a_n = (n+1)(n+2)$

$a_1 = 1$ $a_1 = \dfrac{1}{3}$ $a_1 = 1$
$a_2 = \dfrac{2}{3}$ $a_2 = \dfrac{1}{5}$ $a_2 = \dfrac{1}{2}$ $a_1 = 6$
$a_3 = \dfrac{1}{2}$ $a_3 = \dfrac{1}{7}$ $a_3 = \dfrac{1}{3}$ $a_2 = 12$
$a_4 = \dfrac{2}{5}$ $a_4 = \dfrac{1}{9}$ $a_4 = \dfrac{1}{4}$ $a_3 = 20$
$a_5 = \dfrac{1}{3}$ $a_5 = \dfrac{1}{11}$ $a_5 = \dfrac{1}{5}$ $a_4 = 30$
                                                                                       $a_5 = 42$

54. $\sum_{k=1}^{8} (k-1)!$ or $\sum_{k=0}^{7} k!$ 55. $\sum_{k=1}^{6} \dfrac{1}{k!}$

Lesson 20   Arithmetic

An **arithmetic sequence** is a sequence in which each term after the first is found by adding the preceding term by a constant. This constant value is called the **common difference**. The sequence 3, 8, 13, 18, … is arithmetic because each term can be found by adding 5 to the preceding term.

In general, the common difference can be found by subtracting two consecutive terms in the sequence $a_1, a_2, a_3, …, a_k, a_{k+1}, …a_n$.

$d = a_{k+1} - a_k$ for $k \geq 1$

Example 1: Determine if the sequence is arithmetic. Find the 6th and 10th terms.

a. $11, 7, 3, -1, …$ **b.** $1, 1, 2, 6, …$

Solution:

a. $d = a_2 - a_1 = 7 - 11 = -4$ and $d = a_3 - a_2 = 3 - 7 = -4$

Since d is common, the sequence is arithmetic. To find the 6th term, add the common difference.

$a_5 = a_4 + (-4) = -1 + (-4) = -5$

$a_6 = a_5 + (-4) = -5 + (-4) = -9$

Continue to add –4 to find the 10th term. $a_{10} = -25$

---

b. $d = a_2 - a_1 = 1 - 1 = 0$ and $d = a_3 - a_2 = 2 - 1 = 1$

Since the difference is not common, the sequence is not arithmetic. To find the 6th and 10th terms, you need to establish a pattern. The numbers in the sequence are factorials. Factorials were discussed in Lesson 19.

$a_1 = 0! = 1$ $a_2 = 1! = 1$ $a_3 = 2! = 2$ $a_4 = 3! = 6$

The general term is $a_n = (n-1)!$. Therefore, we can find the 6th and 10th term using the same pattern.

$a_6 = 5! = 120$ $a_{10} = 9! = 362880$

We need to find the general term of the arithmetic sequence. Take a look at the pattern when adding the common difference to the preceding term.

1st Term: a_1

2nd Term: $a_2 = a_1 + d$

3rd Term: $a_3 = a_2 + d = a_1 + d + d = a_1 + 2d$

4th Term: $a_4 = a_3 + d = a_1 + 2d + d = a_1 + 3d$

For any natural number k, $a_k = a_1 + (k-1)d$.

The General Term for any Arithmetic Sequence

The nth term of an arithmetic sequence is given by the formula $a_n = a_1 + (n-1)d$.

Example 2: Find the general term of the sequence. Then find the 25th term.

a. $-12, -6, 0, \ldots$ b. $5, \dfrac{9}{2}, 4, \dfrac{7}{2}, \ldots$

Solution:

a. $a_1 = -12$ $d = a_2 - a_1 = -6 - (-12) = 6$ $a_n = -12 + (n-1)6$ $a_{25} = 6(25) - 18 = 132$

   Common Difference

   $d = a_3 - a_2 = 0 - (-6) = 6$ $a_n = 6n - 18$

---

b. $a_1 = 5$ $d = a_2 - a_1 = \dfrac{9}{2} - 5 = -\dfrac{1}{2}$ $a_n = 5 + (n-1)\left(-\dfrac{1}{2}\right)$ $a_{25} = -\dfrac{1}{2}(25) + \dfrac{11}{2}$

   Common Difference $= -7$

   $d = a_3 - a_2 = 4 - \dfrac{9}{2} = -\dfrac{1}{2}$ $a_n = -\dfrac{1}{2}n + \dfrac{11}{2}$

Using the graphing calculator, we can plot the terms of the sequence. Make sure your calculator is in SEQ Mode. Press Y= and enter the general term $-\dfrac{1}{2}n + \dfrac{11}{2}$.

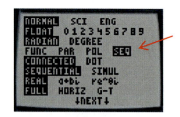
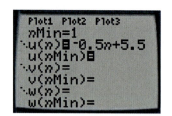

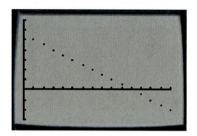

From the points on the graph, we can determine an arithmetic sequence is linear. The common difference is the slope of the linear function. If $d<1$, the terms of the arithmetic sequence will decrease. If $d>1$, the terms of the arithmetic sequence will increase.

Example 3: Find the number of terms in the sequence.

a. $a_1 = 38$, $d = -5$, and $a_n = -27$ **b.** $6, 10, 14, \ldots$ and $a_n = 98$

Solution: Using the general term formula for an arithmetic sequence, substitute the known values and solve for n.

a. $a_1 = 38$ and $d = -5$

$-27 = 38 + (n-1)(-5)$

$-27 = -5n + 43$

$-70 = -5n$

$n = 14$

b. $a_1 = 6$ and $d = 4$

$98 = 6 + (n-1)4$

$98 = 4n + 2$

$96 = 4n$

$n = 24$

Be sure to check your answer by evaluating the general term for the value of n.

Example 4: For the arithmetic sequence, find a_1 and d if $a_6 = -14$ and $a_8 = -20$. Then find the 30th term.

Solution: Use the general term of a arithmetic sequence, $a_n = a_1 + (n-1)d$.

For $n = 6$, $\begin{array}{l} a_6 = a_1 + (6-1)d \\ -14 = a_1 + 5d \end{array}$. For $n = 8$, $\begin{array}{l} a_8 = a_1 + (8-1)d \\ -20 = a_1 + 7d \end{array}$.

Solve the system of equations using the addition method. $\begin{array}{l} -14 = a_1 + 5d \\ -20 = a_1 + 7d \end{array}$

$$-(-14 = a_1 + 5d) \quad \rightarrow \quad 14 = -a_1 - 5d$$
$$-20 = a_1 + 7d \quad \quad \underline{-20 = a_1 + 7d}$$
$$-6 = 2d$$
$$d = -3$$

Use equation 1 to find the value of a_1.
$$-14 = a_1 + 5(-3)$$
$$a_1 = 1$$

The general term is $a_n = 1 + (n-1)(-3)$. To find the 30th term, evaluate the formula for $n = 30$.
$$a_{30} = 1 + (30-1)(-3) = -86$$

Instead of using a system of equations, you can adjust the general term to include the appropriate number multiple of d. For the 8th term, we can write the formula as $a_8 = a_6 + (2)d$ since there are two common differences between terms 6 and 8. Since $a_6 = -14$ and $a_8 = -20$, we can solve for d.

$$a_8 = a_6 + (2)d$$
$$-20 = -14 + (2)d$$
$$d = -3$$

Now, use the regular general term to find $a_1 = 1$.

If we add the terms of an arithmetic sequence, we will find an arithmetic series.

$$S_1 = a_1$$

$$S_2 = a_1 + a_2 = a_1 + (a_1 + d) = 2a_1 + d$$

$$S_3 = a_1 + a_2 + a_3 = a_1 + (a_1 + d) + (a_1 + 2d) = 3a_1 + 3d$$

$$S_4 = a_1 + a_2 + a_3 + a_4 = a_1 + (a_1 + d) + (a_1 + 2d) + (a_1 + 3d) = 4a_1 + 6d$$

$$S_5 = a_1 + a_2 + a_3 + a_4 + a_5 = a_1 + (a_1 + d) + (a_1 + 2d) + (a_1 + 3d) + (a_1 + 4d) = 5a_1 + 10d$$

In general, the partial sum is given by $S_n = a_1 + (a_1 + d) + (a_1 + 2d) + ... + a_n = na_1 + \dfrac{n(n-1)}{2}d$. We can simplify this formula by factoring out $n/2$ from the binomial expression.

$$S_n = na_1 + \frac{n(n-1)}{2}d = \frac{n}{2}[2a_1 + (n-1)d]$$

Note: $2a_1 = a_1 + a_1$

$$= \frac{n}{2}[\underbrace{a_1 + a_1 + (n-1)d}_{a_n}]$$

Remember the general term:
$a_n = a_1 + (n-1)d$

Partial Sum of an Arithmetic Series

The sum of the first n terms of an arithmetic series is given by the formula

$$S_n = \frac{n}{2}(a_1 + a_n)$$

Proof: The sum of an arithmetic series is given by
$S_n = a_1 + (a_1 + d) + (a_1 + 2d) + \ldots + (a_n - 2d) + (a_n - d) + a_n.$
Since addition is commutative, the order in which we add is not important. Writing the series in reverse order gives us $S_n = a_n + (a_n - d) + (a_n - 2d) + \ldots + (a_1 + 2d) + (a_1 + d) + a_1.$

$2S_n = S_n + S_n$

$= a_1 + (a_1 + d) + (a_1 + 2d) + \ldots + (a_n - 2d) + (a_n - d) + a_n$
$+ a_n + (a_n - d) + (a_n - 2d) + \ldots + (a_1 + 2d) + (a_1 + d) + a_1$

If we add corresponding terms, the d variable cancels.

$2S_n = (a_1 + a_n) + (a_1 + a_n) + (a_1 + a_n) + \ldots + (a_1 + a_n) + (a_1 + a_n) + (a_1 + a_n)$

Since $(a_1 + a_n)$ is added n times, it follows that

$2S_n = n(a_1 + a_n)$

$S_n = \frac{n}{2}(a_1 + a_n)$

Example 5: **Find the sum of the first 100 terms of the sequence.**

a. 1, 2, 3, 4, …   b. 0.4, 0.65, 0.9 …

Solution: First, we need to make sure the sequence is arithmetic by finding the common difference.

a. Since the sequence is arithmetic, we can use the arithmetic series formula to find the sum of the first 100 terms. Use $a_1 = 1$ and $d = 1$ to find the 100th term of the sequence.
$a_{100} = 1 + (100 - 1)(1) = 100$
$S_{100} = \frac{100}{2}(1 + 100) = 5050$

b. Use $a_1 = 0.4$ and $d = 0.25$ to find the 100th term of the sequence.
$a_{100} = 0.4 + (100 - 1)(0.25) = 25.15$

$$S_{100} = \frac{100}{2}(0.4 + 25.15) = 1277.5$$

Adding 100 numbers is not difficult. First, you have to find all 100 numbers. Then, add! However, notice the efficiency and accuracy of our work when formulas are used to find the sum.

Example 6: Evaluate the series.

a. $\sum_{k=1}^{30} 2k - 1$
b. $\sum_{k=8}^{20} 7 - 2n$

Solution:

a. First, notice the general term is written as a linear function. Therefore, this series is arithmetic. Use the formula, $S_{30} = \frac{30}{2}(a_1 + a_{30})$. Evaluate using $a_1 = 2(1) - 1 = 1$ and $a_{30} = 2(30) - 1 = 59$.

$$S_{30} = \frac{30}{2}(1 + 59) = 900$$

b. Notice, the first value is $k = 8$. We need to find the sum $a_8 + a_9 + a_{10} + ... + a_{20}$. Therefore, we are finding $S_{20} - S_7$.

$$S_{20} = \frac{20}{2}(a_1 + a_{20}) \qquad S_7 = \frac{7}{2}(a_1 + a_7)$$

$$= 10[5 + (-33)] \qquad = \frac{7}{2}[5 + (-7)]$$

$$= -280 \qquad = -7$$

$$\sum_{k=8}^{20} 7 - 2n = -280 - (-7) = -273$$

Notice, the series in Example 6b is the sum of 13 terms. We can adapt the formula for the arithmetic series to evaluate this sum for $n = 13$. The 1st term will be a_8 and the 13th term will be a_{20}.

$$S_{13} = \frac{13}{2}(a_8 + a_{20}) = \frac{13}{2}[-9 + (-33)] = -273$$

Lesson 20 — Practice Exercises

In Exercises 1 – 6, write the first five terms of the arithmetic sequence using given information.

1. $a_1 = -12$; $d = 5$
2. $a_1 = 7.5$; $d = -0.5$
3. $a_3 = 6$; $d = 4$
4. $a_n = 4n - 9$
5. $a_n = -n + 10$
6. $a_n = -\dfrac{3}{4}n + 12$

In Exercises 7 – 12, verify the sequence is arithmetic. Identify the first term and the common difference. Write an expression for the general term of the sequence. Then find the 5th and 10th terms of the sequence.

7. $1.1, 2.3, 3.5, 4.7, \ldots$
8. $7, -2, -11, -20, \ldots$
9. $2\pi, 3\pi, 4\pi, 5\pi, \ldots$
10. $15, 8, 1, -6, \ldots$
11. $\dfrac{1}{2}, \dfrac{3}{4}, 1, \dfrac{5}{4}, \ldots$
12. $17, 26, 35, 44, \ldots$

In Exercises 13 – 16, find the first term and the common difference using the given information.

13. $a_5 = 23$ and $a_{11} = 58$
14. $a_7 = -20$ and $a_{17} = -50$
15. $a_4 = \dfrac{67}{4}$ and $a_{24} = \dfrac{87}{4}$
16. $a_8 = 2.2$ and $a_{13} = 3.7$

In Exercises 17 – 20, find the number of terms in each arithmetic sequence.

17. $a_1 = 2$, $d = 3$, and $a_n = 155$
18. $a_1 = 0.4$, $d = 0.25$, and $a_n = 10.9$
19. $-5, -1, 3, 7, \ldots, 95$
20. $-3.4, -1.1, 1.2, 3.5, \ldots, 61$

In Exercises 21 – 26, evaluate the sum of the series.

21. $\sum_{k=1}^{15} 5k + 9$
22. $\sum_{k=1}^{100} k$
23. $\sum_{k=1}^{26} -4k + 10$
24. $\sum_{i=1}^{8} -2i - 7$
25. $\sum_{i=1}^{12} \dfrac{1}{2}i + 6$
26. $\sum_{i=1}^{10} \dfrac{3}{4}i$

In Exercises 27 – 30, write the sum in sigma notation. Then evaluate the sum of the series.

27. $2 + 5 + 8 + 11 + \ldots + 44$
28. $-4 + (-8) + (-12) + (-16) + \ldots + (-240)$
29. $1 + 3 + 5 + 7 + \ldots + 299$
30. $36 + 39 + 42 + 45 + \ldots + 63$

Solutions for Practice Exercises **Lesson 20**

1.
$a_1 = -12$
$a_2 = -7$
$a_3 = -2$
$a_4 = 3$
$a_5 = 8$

2.
$a_1 = 7.5$
$a_2 = 7.0$
$a_3 = 6.5$
$a_4 = 6.0$
$a_5 = 5.5$

3.
$a_1 = -2$
$a_2 = 2$
$a_3 = 6$
$a_4 = 10$
$a_5 = 14$

4.
$a_1 = -5$
$a_2 = -1$
$a_3 = 3$
$a_4 = 7$
$a_5 = 11$

5.
$a_1 = 9$
$a_2 = 8$
$a_3 = 7$
$a_4 = 6$
$a_5 = 5$

6.
$a_1 = \dfrac{45}{4}; a_2 = \dfrac{21}{2}$
$a_3 = \dfrac{39}{4}; a_4 = 9$
$a_5 = \dfrac{33}{4}$

7.
$a_1 = 1.1; d = 1.2$
$a_n = 1.2n - 0.1$
$a_5 = 5.9; a_{10} = 11.9$

8.
$a_1 = 7; d = -9$
$a_n = -9n + 16$
$a_5 = -29; a_{10} = -74$

9.
$a_1 = 2\pi; d = \pi$
$a_n = \pi n + \pi$
$a_5 = 6\pi; a_{10} = 11\pi$

10.
$a_1 = 15; d = -7$
$a_n = -7n + 22$
$a_5 = -13; a_{10} = -48$

11.
$a_1 = \dfrac{1}{2}; d = \dfrac{1}{4}$
$a_n = \dfrac{1}{4}n + \dfrac{1}{4}$
$a_5 = \dfrac{3}{2}; a_{10} = \dfrac{11}{4}$

12.
$a_1 = 17; d = 9$
$a_n = 9n + 8$
$a_5 = 53; a_{10} = 98$

13. $a_1 = -\dfrac{1}{3}; d = \dfrac{35}{6}$

14. $a_1 = -2; d = -3$

15. $a_1 = 16; d = \dfrac{1}{4}$

16. $a_1 = \dfrac{1}{10}; d = \dfrac{3}{10}$

17. $n = 52$

18. $n = 43$

19. $n = 26$

20. $n = 29$

21. $S_{15} = 735$

22. $S_{100} = 5050$

23. $S_{26} = -1144$

24. $S_8 = -128$

25. $S_{12} = 111$

26. $S_{10} = \dfrac{165}{4}$

27. $\sum_{k=1}^{15} 3k - 1 = 345$

28. $\sum_{k=1}^{60} -4k = -7320$

29. $\sum_{k=1}^{150} 2k - 1 = 22500$

30. $\sum_{k=1}^{10} 3k + 33 = 495$

Lesson 21 — Geometric

Suppose you are offered a job that pays a salary of $45,000. Every year you will receive a 3% raise. What is your salary for the 2nd, 3rd, and 4th years on the job? If you retire after 40 years, how much will you earn the final year?

1st Year Salary: 45,000

2nd Year Salary: $45,000 + 45,000(0.03)$ *Factor out the GCF 45,000.* $45000(1 + 0.03)$

$45,000(1.03) = 46,350$ $46,350$

3rd Year Salary: $46,350 + 46,350(0.03)$

$46,350(1.03) = 47,740.50$ $47,740.50$

4th Year Salary: $47,740.50 + 47,740.50(0.03)$

$47,740.50(1.03) = 49,172.715$ $49,172.72$

We need to find the salary for the 40th year. Writing down all previous 39 salaries would eventually get us to the 40th year, but there has to be a more efficient way. Can we find a pattern for these numbers?

$a_1 = 45000$

$a_2 = 45000(1.03)$

$a_3 = 46350(1.03)$
    $= 45000(1.03)(1.03) = 45000(1.03)^2$

$a_4 = 47740.50(1.03)$
    $= 45000(1.03)^2(1.03) = 45000(1.03)^3$

Every year depends on the starting salary of $45,000 and 1.03.

$a_{40} = 45000(1.03)^{39} = 142,516.2142$ The 40th year salary will be $142,516.21.

A **geometric sequence** is a sequence in which each term after the first is found by multiplying the preceding term by a constant. This constant value is called the **common ratio**. Recall that an arithmetic sequence has a common difference.

The sequence 45000, 46350, 47740.5, ... is geometric because each term can be found by multiplying the preceding term by 1.03.

In general, the common ratio can be found by dividing two consecutive terms in the sequence $a_1, a_2, a_3, ..., a_k, a_{k+1}, ... a_n$.

$$r = \frac{a_{k+1}}{a_k} \text{ for } k \geq 1$$

Example 1: Determine if the sequence is geometric. Find the 4th and 5th terms.

a. $4, -2, 1, ...$ b. $1, 4, 9, ...$

Solution:

a. $r = \dfrac{a_2}{a_1} = \dfrac{-2}{4} = -\dfrac{1}{2}$ and $r = \dfrac{a_3}{a_2} = \dfrac{1}{-2} = -\dfrac{1}{2}$

Since the r is common, the sequence is geometric.

To find the 4th term, multiply the common ratio and a_3. $a_4 = -\dfrac{1}{2} a_3 = -\dfrac{1}{2}(1) = -\dfrac{1}{2}$

To find the 5th term, multiply the common ratio and a_4. $a_5 = -\dfrac{1}{2} a_4 = -\dfrac{1}{2}\left(-\dfrac{1}{2}\right) = \dfrac{1}{4}$

b. $r = \dfrac{a_2}{a_1} = \dfrac{4}{1} = 4$ and $r = \dfrac{a_3}{a_2} = \dfrac{9}{4}$

Since the r is not the same, the sequence is not geometric.

To find the 4th term, you need to establish a pattern. The numbers in the sequence are perfect squares. Therefore, $a_4 = 16$ and $a_5 = 25$.

For all sequences, it is important to be able to find a general term that will produce each term in the sequence. For geometric sequences, we can establish a pattern using the common ratio.

If the first term and common ratio are known, the following terms depend on a_1 and r. Take a look at the following pattern.

2nd Term: $a_2 = a_1 \cdot r$

3rd Term: $a_3 = a_2 \cdot r = (a_1 \cdot r) r = a_1 \cdot r^2$

4th Term: $a_4 = a_3 \cdot r = (a_1 \cdot r^2) r = a_1 \cdot r^3$

For any natural number k, $a_k = a_{k-1} \cdot r = (a_1 \cdot r^{k-2})r = a_1 \cdot r^{k-1}$

$$a_{k+1} = a_k \cdot r = (a_1 \cdot r^{k-1})r = a_1 \cdot r^k$$

The General Term for any Geometric Sequence

The nth term of a geometric sequence is given by the formula $a_n = a_1 \cdot r^{n-1}$.

Example 2: Find the general term of the sequence. Find the 6th and 9th terms.

a. $4, \dfrac{8}{3}, \dfrac{16}{9}, \ldots$ b. $2, -2\sqrt{3}, 6, \ldots$

Solution:

a. $a_1 = 4$ $r = \dfrac{a_2}{a_1} = \dfrac{\frac{8}{3}}{4} = \dfrac{2}{3}$ $a_n = 4\left(\dfrac{2}{3}\right)^{n-1}$ $a_6 = 4\left(\dfrac{2}{3}\right)^{6-1} = 4\left(\dfrac{32}{243}\right) = \dfrac{128}{243}$

Common Ratio

$r = \dfrac{a_3}{a_2} = \dfrac{\frac{16}{9}}{\frac{8}{3}} = \dfrac{2}{3}$ $a_9 = 4\left(\dfrac{2}{3}\right)^{9-1} = 4\left(\dfrac{256}{6561}\right) = \dfrac{1024}{6561}$

---

b. $a_1 = 2$ $a_n = 2\left(-\sqrt{3}\right)^{n-1}$

$r = \dfrac{a_2}{a_1} = \dfrac{-2\sqrt{3}}{2} = -\sqrt{3}$ $a_6 = 2\left(-\sqrt{3}\right)^{6-1} = 2\underbrace{\left(-\sqrt{3}\right)\left(-\sqrt{3}\right)}_{3}\underbrace{\left(-\sqrt{3}\right)\left(-\sqrt{3}\right)}_{3}\left(-\sqrt{3}\right)$

Common Ratio $= -18\sqrt{3}$

$r = \dfrac{a_3}{a_2} = \dfrac{6}{-2\sqrt{3}} = \dfrac{-3}{\sqrt{3}} = -\sqrt{3}$ $a_9 = 2\left(-\sqrt{3}\right)^{9-1} = 2(81)$
$= 162$

Using the graphing calculator, we can plot the terms of the sequence. Make sure your calculator is in SEQ mode. Press [Y=] and enter the general term $4\left(\dfrac{2}{3}\right)^{n-1}$.

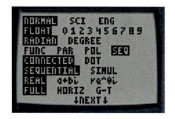

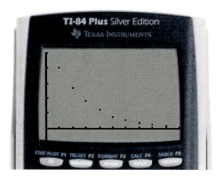

From the points on the graph, we can determine a geometric sequence is exponential. If $|r| > 1$, the geometric sequence follows an exponential growth function. If $0 < |r| < 1$, the geometric sequence follows an exponential decay function.

Example 3: **Find the number of terms in the sequence.**

a. $a_1 = 5$, $r = \sqrt{2}$, and $a_n = 320$

b. $a_1 = 1$, $r = -2$, and $a_n = -128$

Solution: Using the general term formula for a geometric sequence, substitute the known values. Solve for n.

a. $a_n = a_1 \cdot r^{n-1}$

$320 = 5\left(\sqrt{2}\right)^{n-1}$ Divide by a_1

$64 = \left(\sqrt{2}\right)^{n-1}$

$\ln 64 = \ln\left(\sqrt{2}\right)^{n-1}$ Take the logarithm on both sides.

$\ln 64 = (n-1)\ln\left(\sqrt{2}\right)$ Use the Power Rule for logarithms.

$\dfrac{\ln 64}{\ln \sqrt{2}} = n - 1$ Simplify the logarithm.

$12 = n - 1$

$n = 13$

b. $a_n = a_1 \cdot r^{n-1}$

$-128 = 1(-2)^{n-1}$

$128 = (2)^{n-1}$

$\ln 128 = \ln 2^{n-1}$

$\ln 128 = (n-1)\ln 2$

$\dfrac{\ln 128}{\ln 2} = n - 1$

$7 = n - 1$

$n = 8$

Please note: the negatives are not divided here. Since the common ratio is negative, the general term creates an alternating sequence. To find the value of n, we must ignore the negatives when using logarithms to solve the equation.

Remember: the domain of a sequence includes only the natural numbers. If the value of n is a fraction, decimal, or negative number, then a mistake was made in the calculations. Be sure to check your answer by evaluating the general term for the value of n.

Example 4: For the geometric sequence, find a_1 and r if $a_5 = 6$ and $a_9 = 486$.

Solution: Use the general term of a geometric sequence, $a_n = a_1 \cdot r^{n-1}$.

For $n = 5$, $\quad \begin{aligned} a_5 &= a_1 \cdot r^{5-1} \\ 6 &= a_1 \cdot r^4 \end{aligned}$. For $n = 9$, $\quad \begin{aligned} a_9 &= a_1 \cdot r^{9-1} \\ 486 &= a_1 \cdot r^8 \end{aligned}$.

Solve the system of equations using the substitution method. $\quad \begin{aligned} 6 &= a_1 \cdot r^4 \\ 486 &= a_1 \cdot r^8 \end{aligned}$

1st Step: Solve for a variable.

We will use a_1. Equation 1: $\quad a_1 = \dfrac{6}{r^4}$

2nd Step: Substitute for that variable in the other equation.

Equation 2: $\quad 486 = a_1 \cdot r^8$

$$486 = \left(\dfrac{6}{r^4}\right) r^8$$

$$486 = 6r^4 \qquad \text{Divide by 6.}$$

3rd Step: Find the variables.

$$81 = r^4 \qquad \text{Take the 4}^{th}\text{ root on both sides.}$$
Remember: 4th root is the same as an exponent of ¼.

$$(81)^{1/4} = r$$

$$r = 3 \qquad \text{Substitute the value of } r \text{ into the equation.} \quad a_1 = \dfrac{6}{r^4}$$

$$a_1 = \dfrac{6}{3^4} = \dfrac{6}{81} = \dfrac{2}{27}$$

If we add the terms of a geometric sequence, we will find a geometric series.

$$S_1 = a_1$$

$$S_2 = a_1 + a_2 = a_1 + a_1 r$$

$$S_3 = a_1 + a_2 + a_3 = a_1 + a_1 r + a_1 r^2$$

In general, the partial sum is given by $S_n = a_1 + a_1 r + a_1 r^2 + ... + a_1 r^{n-1}$.

If we factor out the GCF a_1, the remaining terms do not combine.

We need to try a different method for finding the formula for the partial sum. Multiply both sides of the series by r, we have $rS_n = a_1 r + a_1 r^2 + a_1 r^3 + ... + a_1 r^n$. Subtract S_n and rS_n.

$$a_1 + a_1 r + a_1 r^2 + ... + a_1 r^{n-1} - (a_1 r + a_1 r^2 + a_1 r^3 + ... + a_1 r^n)$$

Distribute the negative sign, and take a look at the terms that cancel.

$$a_1 + \cancel{a_1 r} + \cancel{a_1 r^2} + ... + \cancel{a_1 r^{n-1}}$$
$$- \cancel{a_1 r} - \cancel{a_1 r^2} - \cancel{a_1 r^3} - .... - \cancel{a_1 r^{n-1}} - a_1 r^n$$

Therefore, $S_n - rS_n = a_1 - a_1 r^n$.

Now, factor out the greatest common factor on both sides and solve for S_n.

$S_n(1-r) = a_1(1-r^n)$ Divide by $1-r$.

$S_n = \dfrac{a_1(1-r^n)}{1-r}$

Partial Sum of a Geometric Series

The sum of the first n terms of a geometric series is given by the formula $S_n = \dfrac{a_1(1-r^n)}{1-r}$.

Note: $r \neq 1$.

Example 5: Find the sum of the first 8 terms of the sequence.

5, 15, 45, ...

Solution: First, we need to determine if the sequence is arithmetic, geometric, or neither. All three sequences have different formulas for finding the series.

Since the difference is not common, this is not arithmetic. Therefore, we will not use the formula found in Lesson 20.

Take a look at the ratio of consecutive terms. $r = \dfrac{a_2}{a_1} = \dfrac{a_3}{a_2} = 3$

Since we have a common ratio, we can use the formula for geometric series.

$$S_8 = \frac{5(1-3^8)}{1-3} = \frac{5(-6560)}{-2} = 16400$$

Example 6: Evaluate the series $\sum_{k=1}^{17} 3\left(\frac{3}{4}\right)^{k-1}$.

Solution: First, we need to determine if the sequence is arithmetic, geometric, or neither. Since the general term is written in geometric form, we can determine that $r = \frac{3}{4}$ and $a_1 = 3$.

$$S_{17} = \frac{3\left(1-\left(\frac{3}{4}\right)^{17}\right)}{1-\frac{3}{4}} \approx 11.9098$$

Be careful using your calculator to simplify the expression. Many parentheses must be used.

$S_\infty = a_1 + a_1 r + a_1 r^2 + \ldots + a_1 r^{n-1} + \ldots$ is called an **infinite geometric series**. Adding an infinite number of terms is impossible. However, if the sum approaches a specific number as n approaches infinity, the sum is called the **limit** of the series. The sum of an infinite series approaches the value of the limit without exceeding it. This behavior is very similar to how an exponential function approaches an asymptote. If the infinite geometric series has a limit, we call the series **convergent**. If the infinite geometric series does not approach a specific number, we call the series **divergent**. How do you determine if the series converges or diverges? It depends on the value of r.

For $|r| < 1$, the value r^n in the formula will become insignificant. If fact, $r^n \to 0$ as $n \to \infty$. The geometric series formula will have a limit if $|r| < 1$.

For $|r| \geq 1$, the infinite geometric series will diverge.

Sum of an Infinite Geometric Series

For $-1 < r < 1$, the sum of an infinite geometric series is given by the formula $S_\infty = \frac{a_1}{1-r}$.

$\frac{a_1}{1-r}$ is the limit of the series.

Example 7: Determine if the following series converges or diverges. If the series converges, find the limit.

a. $5+15+45+...$ b. $\sum_{k=1}^{\infty} 3\left(\dfrac{3}{4}\right)^{k-1}$

Solution:

a. The value of r is greater than 1. Therefore, the geometric series diverges. The sum of the infinite series does not exist.

---

b. Since $r = \dfrac{3}{4}$ is less than 1, the infinite series will converge.

$$S_{\infty} = \dfrac{3}{1-\dfrac{3}{4}} = \dfrac{3}{\dfrac{1}{4}} = 12$$

Some important questions regarding finances can be answered using geometric series. In college algebra, you learned how to apply the compound interest formula to a principal investment. Let's expand that idea.

Example 8: A new graduate decides to save for a house, and he is able to put $250 every month into an account earning 6% compounded monthly. Determine the general term of the geometric sequence. Find the total amount of the investment after 5 years.

Solution: The formula for finding compound interest is $A = P\left(1+\dfrac{r}{n}\right)^{nt}$. The principal amount is $250 and the interest rate is 6% = 0.06. The number of compoundings would be 12. Substitute the known values into the formula.

$A = 250\left(1+\dfrac{0.06}{12}\right)^{12t} = 250(1.005)^{12t}$

Substitute $t = n/12$.

Let $t = \dfrac{n}{12}$, where n is the number of months of the investment.

$A = 250(1.005)^n$ Therefore, the general term for the geometric sequence is $a_n = 250(1.005)^n$.

The 1st deposit of $250 will earn interest for the entire 5 yrs (60 months). $a_{60} = 250(1.005)^{60}$

The 2nd deposit of $250 will earn interest for 59 months. $a_{59} = 250(1.005)^{59}$

The 3rd deposit of $250 will earn interest for 58 months. $a_{58} = 250(1.005)^{58}$

The last deposit of $250 will earn interest for only 1 month. $a_1 = 250(1.005)^1$

The sum of the sequence would result in the total amount of the investment after 5 years.

$$a_1 + a_2 + a_3 + \ldots + a_{59} + a_{60} = \frac{a_1(1-r^n)}{1-r}$$

$$= \frac{250(1.005)(1-1.005^{60})}{1-1.005}$$

$$= \$17{,}529.72$$

It is highly unlikely the graduate will be able to buy a house for $17,000, but this amount would provide a great down payment.

FINAL EXAM REVIEW

KEEP IT FRESH

Let $A = \begin{bmatrix} 4 & 2 \\ 1 & 3 \end{bmatrix}$, $B = \begin{bmatrix} 1 & 2 & 3 & 4 \\ 0 & 2 & -1 & 6 \end{bmatrix}$, and $C = \begin{bmatrix} 0 & 1 \\ 5 & 6 \end{bmatrix}$.

1. Find the order of each matrix.   2. Find $2A - C$.   3. Find $-B$.

4. Find AB.   5. Find $|C|$.   6. Find C^{-1}.

7. Write a matrix equation equivalent to this given system of equations. Then, solve the equation using the inverse of the coefficient matrix.

$$\begin{cases} x + 2y + 5z = 2 \\ 2x + 3y + 8z = 3 \\ -x + y + 2z = 3 \end{cases}$$

8. Solve the system using Cramer's Rule.

$$\begin{cases} 8x - 2y + 5z = 36 \\ 3x + y - z = 17 \\ 2x - 6y + 4z = -2 \end{cases}$$

9. Find the area of the triangle with vertices $(5,3)$, $(-6,5)$, and $(0,7)$.

Lesson 21   Practice Exercises

In Exercises 1 – 6, write the first five terms of the geometric sequence using given information.

1. $a_1 = -4$; $r = -2$

2. $a_1 = 5.2$; $r = -0.5$

3. $a_3 = 27$; $r = 3$

4. $a_n = 3(2)^{n-1}$

5. $a_n = -\left(\dfrac{1}{5}\right)^{n-1}$

6. $a_n = -1.7(4.5)^{n-1}$

In Exercises 7 – 12, verify the sequence is geometric. Identify the first term and the common ratio. Write an expression for the general term of the sequence. Then find the 5th and 10th terms of the sequence.

7. $-16, -4, -1, -0.25, \ldots$

8. $2, -4, 8, -16, \ldots$

9. $7, \dfrac{7}{3}, \dfrac{7}{9}, \dfrac{7}{27}, \ldots$

10. $25, 15, 9, \dfrac{27}{5}, \ldots$

11. $\dfrac{2}{3}, -\dfrac{4}{9}, \dfrac{8}{27}, -\dfrac{16}{81}, \ldots$

12. $-0.008, 0.04, -0.2, 1.0, \ldots$

In Exercises 13 – 16, find the first term and the common ratio using the given information.

13. $a_3 = 12$ and $a_{10} = -1536$

14. $a_2 = 2$ and $a_9 = \dfrac{1}{64}$

15. $a_4 = 125$ and $a_7 = 15625$

16. $a_3 = 0.01$ and $a_8 = 1000$

In Exercises 17 – 20, find the number of terms in each geometric sequence.

17. $a_1 = 36$, $r = \dfrac{1}{3}$, and $a_n = \dfrac{4}{27}$

18. $a_1 = 2$, $r = -3$, and $a_n = 1458$

19. $3, -6, 12, -24, \ldots, -6144$

20. $\dfrac{5}{81}, \dfrac{5}{27}, \dfrac{5}{9}, \dfrac{5}{3}, \ldots, 135$

In Exercises 21 – 29, evaluate the sum of the series.

21. $\displaystyle\sum_{k=1}^{5} -2\left(\dfrac{1}{4}\right)^{k-1}$

22. $\displaystyle\sum_{k=1}^{12} \dfrac{1}{64}(-4)^{k-1}$

23. $\displaystyle\sum_{k=1}^{9} 100(0.5)^{k-1}$

24. $\displaystyle\sum_{i=1}^{6} 0.01(2)^i$

25. $\displaystyle\sum_{k=0}^{10} 3.2(1.1)^{k-1}$

26. $\displaystyle\sum_{k=0}^{8} 4(-3)^k$

27. $\displaystyle\sum_{i=1}^{\infty} 10\left(\dfrac{1}{2}\right)^{i-1}$

28. $\displaystyle\sum_{k=1}^{\infty} 12\left(\dfrac{2}{3}\right)^{k-1}$

29. $\displaystyle\sum_{i=1}^{\infty} (0.1)^{i-1}$

In Exercises 30 – 35, write the sum in sigma notation. Then evaluate the sum of the series.

30. $4+2+1+...+\dfrac{1}{128}$

31. $\dfrac{2}{27}-\dfrac{2}{9}+\dfrac{2}{3}-2+...-162$

32. $0.1+0.01+0.001+...$

33. $0.3+0.03+0.003+...$

34. $\dfrac{3}{64}-\dfrac{3}{16}+\dfrac{3}{4}-3+...+3072$

35. $625+125+25+...+\dfrac{1}{5}$

In Exercises 36 – 41, determine if the sequence given is arithmetic, geometric, or neither. Write an expression for the general term. Then find a_8 and S_8.

36. $1, 3, 9, 27...$

37. $-1, 4, -9, 16,...$

38. $13, 10, 7, 4,...$

39. $-35, -19, -3, 13,...$

40. $2, 6, 12, 20,...$

41. $20, -5, \dfrac{5}{4}, -\dfrac{5}{16},...$

42. Are you willing to work all day for one penny? Not many people would say yes. What if someone offered you a deal that would double your earnings each day for the month of January? You would receive $0.01 on January 1st, $0.02 on January 2nd, $0.04 on January 3rd, and $0.08 on January 4th. Is this a geometric series? If so, find the general term. Determine the amount of money you would receive on January 15th and 31st. Determine the total amount of money you would receive for working the month of January? Would you take the deal?

43. A job pays a salary of $32,000 the first year. Suppose that during the next 39 years there is an annual salary increase of 5.5%. What is the salary for the 40th year? What would the total salary be over a 40-year period?

Solutions for Practice Exercises   **Lesson 21**

1. $a_1 = -4$; $a_2 = 8$; $a_3 = -16$; $a_4 = 32$; $a_5 = -64$

2. $a_1 = 5.2$; $a_2 = -2.6$; $a_3 = 1.3$; $a_4 = -0.65$; $a_5 = 0.325$

3. $a_1 = 3$; $a_2 = 9$; $a_3 = 27$; $a_4 = 81$; $a_5 = 243$

4. $a_1 = 3$; $a_2 = 6$; $a_3 = 12$; $a_4 = 24$; $a_5 = 48$

5. $a_1 = -1$; $a_2 = -\dfrac{1}{5}$; $a_3 = -\dfrac{1}{25}$; $a_4 = -\dfrac{1}{125}$; $a_5 = -\dfrac{1}{625}$

6. $a_1 = -1.7$; $a_2 = -7.65$; $a_3 = -34.425$; $a_4 = -154.9125$; $a_5 = -697.10625$

7. $a_1 = -16$; $r = \dfrac{1}{4}$; $a_n = -16\left(\dfrac{1}{4}\right)^{n-1}$; $a_5 = -\dfrac{1}{16}$; $a_{10} = -\dfrac{1}{16384}$

8. $a_1 = 2$; $r = -2$; $a_n = 2(-2)^{n-1}$; $a_5 = 32$; $a_{10} = -1024$

9. $a_1 = 7$; $r = \dfrac{1}{3}$; $a_n = 7\left(\dfrac{1}{3}\right)^{n-1}$; $a_5 = \dfrac{7}{81}$; $a_{10} = \dfrac{7}{19683}$

10. $a_1 = 25$; $r = \dfrac{3}{5}$; $a_n = 25\left(\dfrac{3}{5}\right)^{n-1}$; $a_5 = \dfrac{81}{25}$; $a_{10} = \dfrac{19683}{78125}$

11. $a_1 = \dfrac{2}{3}$; $r = -\dfrac{2}{3}$; $a_n = \dfrac{2}{3}\left(-\dfrac{2}{3}\right)^{n-1}$; $a_5 = \dfrac{32}{243}$; $a_{10} = -\dfrac{1024}{59049}$

12. $a_1 = -0.008$; $r = -5$; $a_n = -0.008(-5)^{n-1}$; $a_5 = -5$; $a_{10} = 15625$

13. $a_1 = 3$; $r = -2$ 14. $a_1 = 4$; $r = \dfrac{1}{2}$ 15. $a_1 = 1$; $r = 5$ 16. $a_1 = 0.0001$; $r = 10$

17. $n = 6$ 18. $n = 7$ 19. $n = 12$ 20. $n = 8$

21. $S_5 = -\dfrac{341}{128}$ 22. $S_{12} = -\dfrac{3355443}{64}$ 23. $S_9 = \dfrac{12775}{64}$ 24. $S_6 = \dfrac{63}{50}$

25. $S_{11} \approx 53.9088$ 26. $S_9 = 19684$ 27. $S_\infty = 20$ 28. $S_\infty = 36$

29. $S_\infty = \dfrac{10}{9}$ 30. $\sum\limits_{k=1}^{10} 4\left(\dfrac{1}{2}\right)^{k-1} = \dfrac{1023}{128}$ 31. $\sum\limits_{k=1}^{8} \dfrac{2}{27}(-3)^{k-1} = -\dfrac{3280}{27}$

32. $\sum\limits_{k=1}^{\infty} 0.1(0.1)^{k-1} = \dfrac{1}{9}$ 33. $\sum\limits_{k=1}^{\infty} 0.3(0.1)^{k-1} = \dfrac{1}{3}$ 34. $\sum\limits_{k=1}^{9} \dfrac{3}{64}(-4)^{k-1} = \dfrac{157287}{64}$

35. $\sum\limits_{k=1}^{6} 625\left(\dfrac{1}{5}\right)^{k-1} = \dfrac{3906}{5}$ 36. Geometric $a_n = 3^{n-1}$; $a_8 = 2187$; $S_8 = \dfrac{1(1-3^8)}{1-3} = 3280$

37. Neither $a_n = (-1)^n n^2$; $a_8 = 64$; $S_8 = -1 + 4 + (-9) + 16 + (-25) + 36 + (-49) + 64 = 36$

38. Arithmetic $a_n = -3n + 16$; $a_8 = -8$; $S_8 = \dfrac{8}{2}[13 + (-8)] = 20$

39. Arithmetic $a_n = 16n - 51$; $a_8 = 77$; $S_8 = \dfrac{8}{2}(-35 + 77) = 168$

40. Neither $a_n = n(n+1)$; $a_8 = 72$; $S_8 = 2 + 6 + 12 + 20 + 30 + 42 + 56 + 72 = 240$

41. Geometric $a_n = 20\left(-\dfrac{1}{4}\right)^{n-1}$; $a_8 = -\dfrac{5}{4096}$; $S_8 = \dfrac{20\left(1-\left(-\dfrac{1}{4}\right)^8\right)}{1-\left(-\dfrac{1}{4}\right)} = \dfrac{65535}{4096}$

42. $r = 2$, geometric $a_n = 0.01(2)^{n-1}$; $a_{15} = \$163.84$ and $a_{31} = \$10{,}737{,}418.24$; $S_{31} = \dfrac{0.01(1-2^{31})}{1-2} = \$21{,}474{,}836.47$

43. $a_{40} = \$258{,}223.58$ $S_{40} = \$4{,}371{,}379.65$

FINAL EXAM REVIEW

KEEP IT FRESH

Graph the following.

1. Piecewise $f(x) = \begin{cases} -x^2 + 4, & x \le 2 \\ \dfrac{1}{2}x - 5, & x > 2 \end{cases}$

2. Polynomial $P(x) = 3x^3 - 10x^2 + 4x + 8$

3. Rational $f(x) = \dfrac{x^2 - 5x + 6}{x^2 - x - 2}$

4. Exponential $f(x) = \dfrac{1}{2}e^{x-1}$

Final Exam

5. Logarithmic $f(x) = 4 - \log_2 x$

6. Ellipse $25x^2 + 9y^2 - 50x - 72y = 56$

7. Hyperbola $y^2 - x^2 - 8y - 2x = 129$

8. Parabola $y^2 + 4y - 8x - 20 = 0$

Lesson 22 — Mathematical Induction

How would you prove a statement is false? Some may think to use examples to show the statement does not work for all natural numbers. This example would be considered a **counterexample**.

Take a look at the following statement.

$$0 + 1 + 2 + \ldots + (n-1) = n^2 - n$$

For $n = 1$, the statement seems to be true.
$$0 = 1^2 - 1 \checkmark$$

$$a_1 = S_1$$

$$a_1 + a_2 \stackrel{?}{=} S_2$$

However, you can show the statement is false by looking at $n = 2$.
$$0 + 1 \stackrel{?}{=} 2^2 - 2$$
$$1 \neq 2$$

Therefore, the statement $0 + 1 + 2 + \ldots + (n-1) = n^2 - n$ is not true for all natural numbers. To prove a statement false, only a counterexample is needed.

How would you prove a statement is true for all natural numbers? Using specific examples will not give enough proof for all numbers. In fact, it only proves the statement is true for those specific examples. Is the following statement true?

$$0 + 1 + 2 + \ldots + (n-1) = \frac{1}{2}n(n-1)$$

For $n = 1$, the statement is true. $\quad 0 = \frac{1}{2}(1)(1-1) \checkmark$

For $n = 2$, the statement is true. $\quad 0 + 1 = \frac{1}{2}(2)(2-1) \checkmark$

For $n = 3$, the statement is true. $\quad 0 + 1 + 2 = \frac{1}{2}(3)(3-1) \checkmark$

Just because a formula works for the first few examples, does not indicate the statement is true for all natural numbers. In this section, we will learn how to prove a statement is true for all natural numbers using a technique called **mathematical induction**. Before we actually tackle proofs, we need to understand a few notations.

For the general term $a_n = n - 1$, what does a_4 mean? Recall from the previous lessons, the value 4 is evaluated for the variable n.
$$a_4 = 4 - 1 = 3$$

For mathematical induction, we need to find a_k and a_{k+1}. For the general term $a_n = n - 1$,

$a_k = k - 1$ and $a_{k+1} = k + 1 - 1 = k$.

229

For the partial sum $S_n = \frac{1}{2}n(n-1)$, what does S_4 mean? Recall from the previous lessons, the value 4 is evaluated for the variable n.
$$S_4 = \frac{1}{2}(4)(4-1) = 6$$

For mathematical induction, we will need to find S_k and S_{k+1}.

$$S_k = \frac{1}{2}k(k-1) \text{ and } S_{k+1} = \frac{1}{2}(k+1)(k+1-1) = \frac{1}{2}k(k+1)$$

Every occurrence of n is replaced by the number or expression in the subscript.

Example 1: Find a_k, a_{k+1}, S_k, and S_{k+1} for the following.

a. $a_n = 2n+5$ and $S_n = n(n+6)$
b. $a_n = 3(2^{n-1})$ and $S_n = 3(2^n - 1)$

Solution:

a. $a_k = 2k+5 \quad\quad a_{k+1} = 2(k+1)+5 = 2k+7$

$S_k = k(k+6) \quad\quad S_{k+1} = (k+1)(k+1+6) = (k+1)(k+7)$

b. $a_k = 3(2^{k-1}) \quad\quad a_{k+1} = 3(2^{k+1-1}) = 3(2^k)$

$S_k = 3(2^k - 1) \quad\quad S_{k+1} = 3(2^{k+1} - 1)$

There is one more thing to practice before learning how to write a proof by mathematical induction. We need to see the pattern associated with writing partial sums. Take a look at the following sums.

$S_1 = a_1$

$S_2 = a_1 + a_2$

$S_3 = a_1 + a_2 + a_3$

$S_4 = a_1 + a_2 + a_3 + a_4$

From the pattern, we can notice $S_3 = S_2 + a_3$ and $S_4 = S_3 + a_4$. In other words, the previous partial sum plus the following term results in the next partial sum.

$$S_5 = \underbrace{a_1 + a_2 + a_3 + a_4}_{S_4} + a_5$$

Therefore, $S_5 = S_4 + a_5$.

In general, $S_{k+1} = \underbrace{a_1 + a_2 + a_3 + \ldots + a_k}_{S_k} + a_{k+1}$

Example 2: Verify that $S_{k+1} = S_k + a_{k+1}$.

a. $a_n = 2n + 5$ and $S_n = n(n+6)$

b. $a_n = 3(2^{n-1})$ and $S_n = 3(2^n - 1)$

Solution:

a. $S_k + a_{k+1} = k(k+6) + 2(k+1) + 5$

$= k^2 + 6k + 2k + 7$

$= k^2 + 8k + 7$

$= (k+1)(k+7)$

From the partial sum formula,
$S_{k+1} = (k+1)(k+1+6) = (k+1)(k+7)$ ✓

b. $S_k + a_{k+1} = 3(2^k - 1) + 3(2^{k+1-1})$

$= 3(2^k) - 3 + 3(2^k)$

$= 6(2^k) - 3$ Factor out the GCF of 3.

$= 3(2 \cdot 2^k - 1)$ Add exponents.

$= 3(2^{k+1} - 1)$

From the partial sum formula,
$S_{k+1} = 3(2^{k+1} - 1)$ ✓

We are now ready to prove the statement $0 + 1 + 2 + \ldots + (n-1) = \dfrac{1}{2}n(n-1)$ is true for all natural numbers using mathematics induction.

Mathematical Induction

Let $S_n = a_1 + a_2 + a_3 + \ldots + a_n$. The statement is true for all natural numbers if the two steps are met.   **1. Basis Step:** S_1 is true.

2. Induction Step: The truth of S_k implies S_{k+1}.

Example 3: Prove $0 + 1 + 2 + \ldots + (n-1) = \dfrac{1}{2}n(n-1)$ is true for all natural numbers.

Solution: The statement we are proving true is $S_n : 0+1+2+\ldots+(n-1) = \frac{1}{2}n(n-1)$. The needed components for mathematical induction are S_1, S_k, and S_{k+1}.

$S_1 : \ 0 = \frac{1}{2}(1)(1-1)$

$S_k : \ 0+1+2+\ldots+(k-1) = \frac{1}{2}k(k-1)$

$S_{k+1} : \ 0+1+2+\ldots+(k-1)+(k+1-1) = \frac{1}{2}(k+1)(k+1-1)$

Evaluate the series formula for $n = 1$, $n = k$, and $n = k+1$.

Basis Step: For $n = 1$, the statement is true. $S_1 = a_1$

$$\frac{1}{2}(1)(1-1) = 0 \checkmark$$

Induction Step: Assume the statement is true for some $n = k$, where k is any natural number. This means $S_k = \frac{1}{2}k(k-1)$. Verify the statement is true for $n = k + 1$. In other words, show $S_{k+1} = \frac{1}{2}(k+1)(k+1-1)$.

$S_{k+1} = \underbrace{0+1+2+\ldots+(k-1)} + k$

$\quad = \ S_k \ + a_{k+1}$ *Remember: $S_{k+1} = S_k + a_{k+1}$*

$\quad = \frac{1}{2}k(k-1) + k$ *Substitute for S_k and a_{k+1}.*

$\quad = \frac{1}{2}k^2 - \frac{1}{2}k + k$ *Distribute and collect like terms.*

$\quad = \frac{1}{2}k^2 + \frac{1}{2}k$

$\quad = \frac{1}{2}k(k+1)$ *Factor out the GCF.*

$\quad = \frac{1}{2}(k+1)(k+1-1) \checkmark$ *Note: the result is S_{k+1}*

Since S_k implies S_{k+1}, the statement $0+1+2+\ldots+(n-1) = \frac{1}{2}n(n-1)$ is true for all natural numbers by the principle of mathematical induction.

This proof may seem a little vague, almost as if we did not do anything. However, the key to mathematical induction is the wording. Assuming the statement is true for **any** natural number k, we show that the statement would work for **all** natural numbers. Think of mathematical induction as a domino rally. When the dominoes are lined up, induction gives us two steps. For the basis step, the first domino must be knocked over. For the induction step, if any one domino is knocked over, then the one next to it will also be hit and knocked over. A chain reaction occurs: S_1 implies S_2, S_2 implies S_3, S_3 implies S_4, S_k implies S_{k+1}, and so on. In order for all dominoes to fall, the basis step and the induction step must be satisfied.

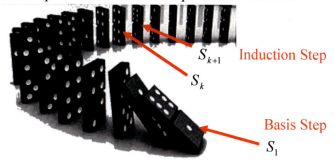

Example 4: Prove $1+2+4+8+...+2^{n-1} = 2^n - 1$ **is true for all natural numbers.**

Solution: The statement we are proving true is $S_n: 1+2+4+8+...+2^{n-1} = 2^n - 1$. The needed components for mathematical induction are S_1, S_k, and S_{k+1}.

$S_1: \ 1 = 2^1 - 1$

$S_k: \ 1+2+4+8+...+2^{k-1} = 2^k - 1$

$S_{k+1}: \ 1+2+4+8+...+2^{k-1} + 2^{k+1-1} = 2^{k+1} - 1$

Basis Step: For $n = 1$, the statement is true. $\quad S_1 = a_1$
$$2^1 - 1 = 1 \checkmark$$

Induction Step: Assume the statement is true for some $n = k$, where k is any natural number. This means $S_k = 2^k - 1$. Verify the statement is true for $n = k + 1$. In other words, show $S_{k+1} = 2^{k+1} - 1$.

$S_{k+1} = \underbrace{1+2+4+8+...+2^{k-1}}_{} + 2^{k+1-1}$

$\quad = \ S_k \ + a_{k+1}$ Remember: $S_{k+1} = S_k + a_{k+1}$

$\quad = 2^k - 1 + 2^k$ Substitute for S_k and a_{k+1}.

$\quad = 2 \cdot 2^k - 1$ Add like terms: $2^k + 2^k = 2(2^k)$

$\quad = 2^{k+1} - 1 \checkmark$ When multiplying like bases, add exponents.

Since S_k implies S_{k+1}, the statement $1+2+4+8+...+2^{n-1} = 2^n - 1$ is true for all natural numbers by the principle of mathematical induction.

In calculus, we will need to show a statement that includes an inequality is true for all natural numbers. Mathematical Induction can be used for these statements as well.

233

Example 5: Prove $3^n < 3^{n+1}$ is true for all natural numbers.

Solution: The statement we are proving true is $S_n : 3^n < 3^{n+1}$. The needed components for mathematical induction are $S_1 : 3^1 < 3^{1+1}, S_k : 3^k < 3^{k+1}$, and $S_{k+1} : 3^{k+1} < 3^{k+1+1}$.

Basis Step: For $n = 1$, the statement is true.
$$3^1 < 3^{1+1}$$
$$3 < 9 \checkmark$$

Induction Step: Assume the statement is true for some $n = k$, where k is any natural number. This means $3^k < 3^{k+1}$. Since the statement is assumed true, we can use it to show the statement is true for $n = k + 1$.

Multiply both sides by 3. $\quad 3 \cdot 3^k < 3 \cdot 3^{k+1}$

When multiplying, the exponents are added. $\quad 3^{k+1} < 3^{k+2}$

Since S_k implies S_{k+1}, the statement $3^n < 3^{n+1}$ is true for all natural numbers by the principle of mathematical induction.

FINAL EXAM REVIEW

KEEP IT FRESH

1. Write the ellipse in standard form. Find the center, vertices, and the foci.

$$9x^2 + 4y^2 - 18x + 16y - 11 = 0$$

2. Find the center, vertices, foci, and the asymptotes. Then graph.

$$\frac{(x-2)^2}{9} - \frac{(y+1)^2}{16} = 1$$

3. Find the equation of a parabola with directrix $x = -5$ and focus $(5, 0)$.

Lesson 22 — Practice Exercises

In Exercises 1 – 4, find a_k and a_{k+1}.

1. $a_n = 4n - 1$
2. $a_n = 2^n$
3. $a_n = n^2$
4. $a_n = \dfrac{1}{n(n+1)}$

In Exercises 5 – 8, find S_k and S_{k+1}. Verify that $S_{k+1} = S_k + a_{k+1}$.

5. $a_n = 4n - 1$
6. $a_n = 2^n$
7. $a_n = n^2$
8. $a_n = \dfrac{1}{n(n+1)}$

$S_n = n(2n+1)$ $\qquad$ $S_n = 2(2^n - 1)$ $\qquad$ $S_n = \dfrac{n(n+1)(2n+1)}{6}$ $\qquad$ $S_n = \dfrac{n}{n+1}$

In Exercises 9 – 20, use mathematical induction to prove the indicated formula for the series is true for all natural numbers.

9. $3 + 7 + 11 + \ldots + (4n - 1) = n(2n + 1)$

10. $2 + 4 + 8 + \ldots + 2^n = 2(2^n - 1)$

11. $1 + 4 + 9 + \ldots + n^2 = \dfrac{n(n+1)(2n+1)}{6}$

12. $\dfrac{1}{2} + \dfrac{1}{6} + \dfrac{1}{12} + \ldots + \dfrac{1}{n(n+1)} = \dfrac{n}{n+1}$

13. $2 + 6 + 18 + \ldots + 2(3^{n-1}) = 3^n - 1$

14. $10 + 17 + 24 + \ldots + (7n + 3) = \dfrac{n(7n + 13)}{2}$

15. $\dfrac{1}{3} + \dfrac{1}{15} + \dfrac{1}{35} + \ldots + \dfrac{1}{(2n-1)(2n+1)} = \dfrac{n}{2n+1}$

16. $1 + 8 + 27 + \ldots + n^3 = \dfrac{n^2(n+1)^2}{4}$

17. $1 + 3 + 5 + \ldots + (2n - 1) = n^2$

18. $4 + 16 + 64 + \ldots + (4^n) = \dfrac{4}{3}(4^n - 1)$

19. $1 + 4 + 7 + \ldots + (3n - 2) = \dfrac{n(3n-1)}{2}$

20. $2 + 6 + 12 + \ldots + n(n+1) = \dfrac{n(n+1)(n+2)}{3}$

In Exercises 21 – 25, use mathematical induction to prove the statement is true for all natural numbers.

21. $n < 2^n$
22. $4^n < 4^{n+1}$
23. $n < n + 1$
24. $\log n < n$
25. $2n \leq 2^n$

235

Solutions for Practice Exercises Lesson 22

1. $a_k = 4k - 1$
2. $a_k = 2^k$
3. $a_k = k^2$
4. $a_k = \dfrac{1}{k(k+1)}$

   $a_{k+1} = 4k + 3$
   $a_{k+1} = 2^{k+1}$
   $a_{k+1} = (k+1)^2$
   $a_{k+1} = \dfrac{1}{(k+1)(k+2)}$

5. $S_k = k(2k+1)$

   $S_{k+1} = (k+1)(2k+3)$

   $S_k + a_{k+1} = k(2k+1) + 4(k+1) - 1$
   $= 2k^2 + k + 4k + 4 - 1$
   $= 2k^2 + 5k + 3$
   $= (k+1)(2k+3)$

6. $S_k = 2(2^k - 1)$

   $S_{k+1} = 2(2^{k+1} - 1)$

   $S_k + a_{k+1} = 2(2^k - 1) + 2^{k+1}$
   $= 2^{k+1} - 2 + 2^{k+1}$
   $= 2(2^{k+1}) - 2$
   $= 2(2^{k+1} - 1)$

7. $S_k = \dfrac{k(k+1)(2k+1)}{6}$

   $S_{k+1} = \dfrac{(k+1)(k+2)(2k+3)}{6}$

   $S_k + a_{k+1} = \dfrac{k(k+1)(2k+1)}{6} + (k+1)^2$
   $= \dfrac{k(k+1)(2k+1)}{6} + \dfrac{6(k+1)^2}{6}$
   $= \dfrac{2k^3 + 3k^2 + k}{6} + \dfrac{6k^2 + 12k + 6}{6}$
   $= \dfrac{2k^3 + 9k^2 + 13k + 6}{6}$
   $= \dfrac{(k+1)(k+2)(2k+3)}{6}$

8. $S_k = \dfrac{k}{k+1}$

   $S_{k+1} = \dfrac{k+1}{k+2}$

   $S_k + a_{k+1} = \dfrac{k}{k+1} + \dfrac{1}{(k+1)(k+2)}$
   $= \dfrac{k(k+2)}{(k+1)(k+2)} + \dfrac{1}{(k+1)(k+2)}$
   $= \dfrac{k^2 + 2k + 1}{(k+1)(k+2)}$
   $= \dfrac{(k+1)(k+1)}{(k+1)(k+2)}$
   $= \dfrac{k+1}{k+2}$

9. The statement we are proving true is $3+7+11+\ldots+(4n-1)=n(2n+1)$. The needed components for mathematical induction are $S_1: 3=1(2\cdot 1+1)$, $S_k: 3+7+11+\ldots+(4k-1)=k(2k+1)$, and $S_{k+1}: 3+7+11+\ldots+(4k-1)+(4(k+1)-1)=(k+1)(2(k+1)+1)$.

Basis Step: For $n=1$, the statement is true. $S_1 = a_1$
$$1(2\cdot 1+1)=3$$

Induction Step: Assume the statement is true for some $n = k$, where k is any natural number. This means $S_k = k(2k+1)$. Verify the statement is true for $n = k + 1$. In other words, show $S_{k+1} = (k+1)(2k+3)$.

$$S_{k+1} = 3+7+1+\ldots+(4k-1)+(4(k+1)-1)$$
$$= S_k + a_{k+1}$$
$$= k(2k+1)+4(k+1)-1$$
$$= 2k^2+k+4k+4-1$$
$$= 2k^2+5k+3$$
$$= (k+1)(2k+3) \checkmark$$

Since S_k implies S_{k+1}, the statement $3+7+11+\ldots+(4n-1)=n(2n+1)$ is true for all natural numbers by the principle of mathematical induction.

10. The statement we are proving true is $2+4+8+\ldots+2^n = 2(2^n-1)$. The needed components for mathematical induction are $S_1: 2=2(2^1-1)$, $S_k: 2+4+8+\ldots+2^k = 2(2^k-1)$, and $S_{k+1}: 2+4+8+\ldots+2^k+2^{k+1} = 2(2^{k+1}-1)$.

Basis Step: For $n = 1$, the statement is true. $S_1 = a_1$
$$2(2^1-1) = 2$$

Induction Step: Assume the statement is true for some $n = k$, where k is any natural number. This means $S_k = 2(2^k-1)$. Verify the statement is true for $n = k + 1$. In other words, show $S_{k+1} = 2(2^{k+1}-1)$.

$$S_{k+1} = 2+4+8+\ldots+2^k+2^{k+1}$$
$$= S_k + a_{k+1}$$
$$= 2(2^k-1)+2^{k+1}$$
$$= 2^{k+1}-2+2^{k+1}$$
$$= 2\cdot 2^{k+1}-2$$
$$= 2(2^{k+1}-1) \checkmark$$

Since S_k implies S_{k+1}, the statement $2+4+8+\ldots+2^n = 2(2^n-1)$ is true for all natural numbers by the principle of mathematical induction.

11. The statement we are proving true is $1+4+9+\ldots+n^2 = \dfrac{n(n+1)(2n+1)}{6}$. The needed components for mathematical induction are $S_1: 1=\dfrac{1(1+1)(2\cdot 1+1)}{6}$, $S_k: 1+4+9+\ldots+k^2 = \dfrac{k(k+1)(2k+1)}{6}$, and $S_{k+1}: 1+4+9+\ldots k^2+(k+1)^2 = \dfrac{(k+1)(k+1+1)(2(k+1)+1)}{6} = \dfrac{(k+1)(k+2)(2k+3)}{6}$.

Basis Step: For $n = 1$, the statement is true. $S_1 = a_1$
$$\frac{1(1+1)(2\cdot 1+1)}{6} = 1$$

Induction Step: Assume the statement is true for some $n = k$, where k is any natural number. This means $S_k = \dfrac{k(k+1)(2k+1)}{6}$. Verify the statement is true for $n = k + 1$. In other words, show $S_{k+1} = \dfrac{(k+1)(k+2)(2k+3)}{6}$.

$$S_{k+1} = 1+4+9+\ldots+k^2+(k+1)^2$$
$$= S_k + a_{k+1}$$
$$= \frac{k(k+1)(2k+1)}{6}+(k+1)^2$$
$$= \frac{k(k+1)(2k+1)}{6}+\frac{6(k+1)^2}{6}$$
$$= \frac{(k+1)[k(2k+1)+6(k+1)]}{6}$$
$$= \frac{(k+1)[2k^2+7k+6]}{6}$$
$$= \frac{(k+1)(k+2)(2k+3)}{6} \checkmark$$

Since S_k implies S_{k+1}, the statement $1+4+9+\ldots+n^2 = \dfrac{n(n+1)(2n+1)}{6}$ is true for all natural numbers by the principle of mathematical induction.

12. The statement we are proving true is $\frac{1}{2}+\frac{1}{6}+\frac{1}{12}+\ldots+\frac{1}{n(n+1)}=\frac{n}{n+1}$. The needed components for mathematical induction are $S_1: \frac{1}{2}=\frac{1}{1+1}$, $S_k: \frac{1}{2}+\frac{1}{6}+\frac{1}{12}+\ldots+\frac{1}{k(k+1)}=\frac{k}{k+1}$, and $S_{k+1}: \frac{1}{2}+\frac{1}{6}+\frac{1}{12}+\ldots\frac{1}{k(k+1)}+\frac{1}{(k+1)(k+1+1)}=\frac{k+1}{k+1+1}$.

Basis Step: For $n = 1$, the statement is true. $S_1 = a_1$
$$\frac{1}{1+1}=\frac{1}{2}$$

Induction Step: Assume the statement is true for some $n = k$, where k is any natural number. This means $S_k = \frac{k}{k+1}$. Verify the statement is true for $n = k + 1$. In other words, show $S_{k+1}=\frac{k+1}{k+1+1}=\frac{k+1}{k+2}$.

$$S_{k+1} = \frac{1}{2}+\frac{1}{6}+\frac{1}{12}+\ldots+\frac{1}{k(k+1)}+\frac{1}{(k+1)(k+1+1)}$$
$$= S_k + a_{k+1}$$
$$= \frac{k}{k+1}+\frac{1}{(k+1)(k+2)}$$
$$= \frac{k(k+2)}{(k+1)(k+2)}+\frac{1}{(k+1)(k+2)}$$
$$= \frac{k(k+2)+1}{(k+1)(k+2)}$$
$$= \frac{k^2+2k+1}{(k+1)(k+2)}$$
$$= \frac{(k+1)^2}{(k+1)(k+2)}$$
$$= \frac{k+1}{k+2} \checkmark$$

Since S_k implies S_{k+1}, the statement $\frac{1}{2}+\frac{1}{6}+\frac{1}{12}+\ldots+\frac{1}{n(n+1)}=\frac{n}{n+1}$ is true for all natural numbers by the principle of mathematical induction.

13. The statement we are proving true is $2+6+18+\ldots+2(3^{n-1})=3^n-1$. The needed components for mathematical induction are $S_1: 2=3^1-1$, $S_k: 2+6+18+\ldots+2(3^{k-1})=3^k-1$, and $S_{k+1}: 2+6+18+\ldots 2(3^{k-1})+2(3^{k+1-1})=3^{k+1}-1$.

Basis Step: For $n = 1$, the statement is true. $S_1 = a_1$
$$3^1-1=2$$

Induction Step: Assume the statement is true for some $n = k$, where k is any natural number. This means $S_k = 3^k-1$. Verify the statement is true for $n = k + 1$. In other words, show $S_{k+1}=3^{k+1}-1$.

$$S_{k+1} = 2+6+18+\ldots 2(3^{k-1})+2(3^{k+1-1})$$
$$= S_k + a_{k+1}$$
$$= 3^k-1+2(3^{k+1-1})$$
$$= 3^k-1+2(3^k)$$
$$= 3(3^k)-1$$
$$= 3^{k+1}-1 \checkmark$$

Since S_k implies S_{k+1}, the statement $2+6+18+\ldots+2(3^{n-1})=3^n-1$ is true for all natural numbers by the principle of mathematical induction.

14. The statement we are proving true is $10+17+24+\ldots+(7n+3)=\frac{n(7n+13)}{2}$. The needed components for mathematical induction are $S_1: 7(1)+3=\frac{1(7\cdot 1+13)}{2}$, $S_k: 10+17+24+\ldots+(7k+3)=\frac{k(7k+13)}{2}$, and $S_{k+1}: 10+17+24+\ldots+(7k+3)+(7(k+1)+3)=\frac{(k+1)(7(k+1)+13)}{2}$.

Basis Step: For $n = 1$, the statement is true. $S_1 = a_1$
$$\frac{1(7\cdot 1+13)}{2}=10$$

Induction Step: Assume the statement is true for some $n = k$, where k is any natural number. This means $S_k = \frac{k(7k+13)}{2}$. Verify the statement is true for $n = k + 1$. In other words, show $S_{k+1}=\frac{(k+1)(7(k+1)+13)}{2}=\frac{(k+1)(7k+20)}{2}$.

$$S_{k+1} = 10+17+24+\ldots+(7k+3)+(7(k+1)+3)$$
$$= S_k + a_{k+1}$$

$$= \frac{k(7k+13)}{2} + (7(k+1)+3)$$

$$= \frac{k(7k+13)}{2} + (7k+10)\frac{2}{2}$$

$$= \frac{k(7k+13)}{2} + \frac{2(7k+10)}{2}$$

$$= \frac{7k^2+13k+14k+20}{2}$$

$$= \frac{7k^2+27k+20}{2}$$

$$= \frac{(k+1)(7k+20)}{2} \checkmark$$

Since S_k implies S_{k+1}, the statement $10+17+24+\ldots+(7n+3) = \frac{n(7n+13)}{2}$ is true for all natural numbers by the principle of mathematical induction.

15. The statement we are proving true is $\frac{1}{3}+\frac{1}{15}+\frac{1}{35}+\ldots+\frac{1}{(2n-1)(2n+1)} = \frac{n}{2n+1}$. The needed components for mathematical induction are:

$S_1: \frac{1}{(2\cdot1-1)(2\cdot1+1)} = \frac{1}{2\cdot1+1}$, $S_k: \frac{1}{3}+\frac{1}{15}+\frac{1}{35}+\ldots+\frac{1}{(2k-1)(2k+1)} = \frac{k}{2k+1}$, and

$S_{k+1}: \frac{1}{3}+\frac{1}{15}+\frac{1}{35}+\ldots+\frac{1}{(2k-1)(2k+1)} + \frac{1}{(2(k+1)-1)(2(k+1)+1)} = \frac{k+1}{2(k+1)+1}$.

Basis Step: For $n = 1$, the statement is true. $S_1 = a_1$

$$\frac{1}{(2\cdot1-1)(2\cdot1+1)} = \frac{1}{3}$$

Induction Step: Assume the statement is true for some $n = k$, where k is any natural number. This means $S_k = \frac{k}{2k+1}$. Verify the statement is true for $n = k+1$. In other words, show $S_{k+1} = \frac{k+1}{2(k+1)+1} = \frac{k+1}{2k+3}$.

$$S_{k+1} = \frac{1}{3}+\frac{1}{15}+\frac{1}{35}+\ldots+\frac{1}{(2k-1)(2k+1)} + \frac{1}{(2(k+1)-1)(2(k+1)+1)}$$

$$= S_k + a_{k+1}$$

$$= \frac{k}{2k+1} + \frac{1}{(2(k+1)-1)(2(k+1)+1)}$$

$$= \frac{k}{2k+1} + \frac{1}{(2k+1)(2k+3)}$$

$$= \frac{k}{2k+1}\left(\frac{2k+3}{2k+3}\right) + \frac{1}{(2k+1)(2k+3)}$$

$$= \frac{k(2k+3)}{(2k+1)(2k+3)} + \frac{1}{(2k+1)(2k+3)}$$

$$= \frac{2k^2+3k+1}{(2k+1)(2k+3)}$$

$$= \frac{(2k+1)(k+1)}{(2k+1)(2k+3)}$$

$$= \frac{k+1}{2k+3} \checkmark$$

Since S_k implies S_{k+1}, the statement $\frac{1}{3}+\frac{1}{15}+\frac{1}{35}+\ldots+\frac{1}{(2n-1)(2n+1)} = \frac{n}{2n+1}$ is true for all natural numbers by the principle of mathematical induction.

16. The statement we are proving true is $1+8+27+\ldots+n^3 = \frac{n^2(n+1)^2}{4}$. The needed components for mathematical induction are: $S_1: n^3 = \frac{n^2(n+1)^2}{4}$,

$S_k: 1+8+27+\ldots+k^3 = \frac{k^2(k+1)^2}{4}$, and $S_{k+1}: 1+8+27+\ldots+k^3 + (k+1)^3 = \frac{(k+1)^2(k+1+1)^2}{4}$.

Basis Step: For $n = 1$, the statement is true. $S_1 = a_1$

$$\frac{1^2(1+1)^2}{4} = 1$$

Induction Step: Assume the statement is true for some $n = k$, where k is any natural number. This means $S_k = \frac{k^2(k+1)^2}{4}$. Verify the statement is true for $n = k+1$. In other words, show $S_{k+1} = \frac{(k+1)^2(k+1+1)^2}{4} = \frac{(k+1)^2(k+2)^2}{4}$.

239

$S_{k+1} = 1+8+27+...+k^3 +(k+1)^3$

$\phantom{S_{k+1}} = S_k + a_{k+1}$

$\phantom{S_{k+1}} = \dfrac{k^2(k+1)^2}{4} + (k+1)^3$

$\phantom{S_{k+1}} = \dfrac{k^2(k+1)^2}{4} + \dfrac{4(k+1)^3}{4}$

$\phantom{S_{k+1}} = \dfrac{k^2(k+1)^2 + 4(k+1)^3}{4}$

$\phantom{S_{k+1}} = \dfrac{(k+1)^2[k^2 + 4(k+1)]}{4}$

$\phantom{S_{k+1}} = \dfrac{(k+1)^2(k^2 + 4k + 4)]}{4}$

$\phantom{S_{k+1}} = \dfrac{(k+1)^2(k+2)^2}{4}$ ✓

Since S_k implies S_{k+1}, the statement $1+8+27+...+n^3 = \dfrac{n^2(n+1)^2}{4}$ is true for all natural numbers by the principle of mathematical induction.

17. The statement we are proving true is $1+3+5...+(2n-1)=n^2$. The needed components for mathematical induction are $S_1: 2(1)-1=1^2$, $S_k: 1+3+5...+(2k-1)=k^2$, and $S_{k+1}: 1+3+5...+(2k-1)+(2(k+1)-1)=(k+1)^2$.

Basis Step: For $n = 1$, the statement is true. $S_1 = a_1$
$ 1^2 = 1$

Induction Step: Assume the statement is true for some $n = k$, where k is any natural number. This means $S_k = k^2$. Verify the statement is true for $n = k + 1$. In other words, show $S_{k+1} = (k+1)^2$.

$S_{k+1} = 1+3+5...+(2k-1)+(2(k+1)-1)$

$\phantom{S_{k+1}} = S_k + a_{k+1}$

$\phantom{S_{k+1}} = k^2 + (2k+1)$

$\phantom{S_{k+1}} = (k+1)^2$ ✓

Since S_k implies S_{k+1}, the statement $1+3+5...+(2n-1)=n^2$ is true for all natural numbers by the principle of mathematical induction.

18. The statement we are proving true is $4+16+64+...+(4^n) = \dfrac{4}{3}(4^n - 1)$. The needed components for mathematical induction are:

$S_1: 4^1 = \dfrac{4}{3}(4^1 - 1)$; $S_k: 4+16+64+...+(4^k) = \dfrac{4}{3}(4^k - 1)$; and $S_{k+1}: 4+16+64+...+(4^k)+(4^{k+1}) = \dfrac{4}{3}(4^{k+1} - 1)$.

Basis Step: For $n = 1$, the statement is true. $S_1 = a_1$
$ \dfrac{4}{3}(4^1 - 1) = 4$

Induction Step: Assume the statement is true for some $n = k$, where k is any natural number. This means $S_k = \dfrac{4}{3}(4^k - 1)$. Verify the statement is true for $n = k + 1$. In other words, show $S_{k+1} = \dfrac{4}{3}(4^{k+1} - 1)$.

$S_{k+1} = 4+16+64+...+(4^k)+(4^{k+1})$

$\phantom{S_{k+1}} = S_k + a_{k+1}$

$\phantom{S_{k+1}} = \dfrac{4}{3}(4^k - 1) + 4^{k+1}$

$\phantom{S_{k+1}} = \dfrac{4^{k+1}}{3} - \dfrac{4}{3} + 4^{k+1}$

$\phantom{S_{k+1}} = \dfrac{4^{k+1}}{3} - \dfrac{4}{3} + \dfrac{3 \cdot 4^{k+1}}{3}$

$\phantom{S_{k+1}} = \dfrac{4 \cdot 4^{k+1}}{3} - \dfrac{4}{3}$

$\phantom{S_{k+1}} = \dfrac{4}{3}(4^{k+1} - 1)$ ✓

Since S_k implies S_{k+1}, the statement $4+16+64+...+(4^n) = \dfrac{4}{3}(4^n - 1)$ is true for all natural numbers by the principle of mathematical induction.

19. The statement we are proving true is $1+4+7+...+(3n-2) = \dfrac{n(3n-1)}{2}$. The needed components for mathematical induction are:

$S_1: 3(1)-2 = \dfrac{1(3 \cdot 1-1)}{2}$; $S_k: 1+4+7+...+(3k-2) = \dfrac{k(3k-1)}{2}$, and $S_{k+1}: 1+4+7+...+(3k-2)+(3(k+1)-2) = \dfrac{(k+1)(3(k+1)-1)}{2}$.

Basis Step: For $n = 1$, the statement is true. $\quad S_1 = a_1$
$$\frac{1(3 \cdot 1 - 1)}{2} = 1$$

Induction Step: Assume the statement is true for some $n = k$, where k is any natural number. This means $S_k = \frac{k(3k-1)}{2}$. Verify the statement is true for $n = k + 1$. In other words, show $S_{k+1} = \frac{(k+1)(3(k+1)-1)}{2} = \frac{(k+1)(3k+2)}{2}$.

$S_{k+1} = 1 + 4 + 7 + \ldots + (3k-2) + (3(k+1)-2)$
$= S_k + a_{k+1}$
$= \frac{k(3k-1)}{2} + 3k + 1$
$= \frac{k(3k-1)}{2} + \frac{2(3k+1)}{2}$
$= \frac{3k^2 - k + 6k + 2}{2}$
$= \frac{3k^2 + 5k + 2}{2}$
$= \frac{(k+1)(3k+2)}{2}$ ✓

Since S_k implies S_{k+1}, the statement $1 + 4 + 7 + \ldots + (3n-2) = \frac{n(3n-1)}{2}$ is true for all natural numbers by the principle of mathematical induction.

20. The statement we are proving true is $2 + 6 + 12 + \ldots + n(n+1) = \frac{n(n+1)(n+2)}{3}$. The needed components for mathematical induction are:

$S_1: 1(1+1) = \frac{1(1+1)(1+2)}{3}$, $S_k: 2 + 6 + 12 + \ldots + k(k+1) = \frac{k(k+1)(k+2)}{3}$, and $S_{k+1}: 2 + 6 + 12 + \ldots + k(k+1) + (k+1)(k+1+1) = \frac{(k+1)(k+1+1)(k+1+2)}{3}$.

Basis Step: For $n = 1$, the statement is true. $\quad S_1 = a_1$
$$\frac{1(1+1)(1+2)}{3} = 2$$

Induction Step: Assume the statement is true for some $n = k$, where k is any natural number. This means $S_k = \frac{k(k+1)(k+2)}{3}$. Verify the statement is true for $n = k + 1$. In other words, show $S_{k+1} = \frac{(k+1)(k+1+1)(k+1+2)}{3} = \frac{(k+1)(k+2)(k+3)}{3}$.

$S_{k+1} = 2 + 6 + 12 + \ldots + k(k+1) + (k+1)(k+1+1)$
$= S_k + a_{k+1}$
$= \frac{k(k+1)(k+2)}{3} + (k+1)(k+2)$
$= \frac{k(k+1)(k+2)}{3} + \frac{3(k+1)(k+2)}{3}$
$= \frac{k(k+1)(k+2) + 3(k+1)(k+2)}{3}$
$= \frac{(k+1)(k+2)(k+3)}{3}$ ✓

Since S_k implies S_{k+1}, the statement $2 + 6 + 12 + \ldots + n(n+1) = \frac{n(n+1)(n+2)}{3}$ is true for all natural numbers by the principle of mathematical induction.

21. The statement we are proving true is $S_n : n < 2^n$. The needed components for mathematical induction are $S_1 : 1 < 2^1$, $S_k : k < 2^k$, and $S_{k+1} : k+1 < 2^{k+1}$.

Basis Step: For $n = 1$, the statement is true. $\quad 1 < 2^1$ ✓
Induction Step: Assume the statement is true for some $n = k$, where k is any natural number. This means $k < 2^k$. Since the statement is assumed true, we can use it to show the statement is true for $n = k + 1$.
Multiply both sides by 2. $\quad\quad 2k < 2 \cdot 2^k$
When multiplying, the exponents are added. $\quad 2k < 2^{k+1}$
$\quad\quad\quad k + k < 2^{k+1}$
Since k is a natural number, $k + 1 \leq k + k$. Therefore, $k + 1 \leq k + k < 2^{k+1}$
This means $\quad\quad\quad k + 1 < 2^{k+1}$.
Since S_k implies S_{k+1}, the statement $n < 2^n$ is true for all natural numbers by the principle of mathematical induction.

22. The statement we are proving true is $S_n : 4^n < 4^{n+1}$. The needed components for mathematical induction are $S_1 : 4^1 < 4^{1+1}$, $S_k : 4^k < 4^{k+1}$, and $S_{k+1} : 4^{k+1} < 4^{k+1+1}$.

Basis Step: For $n = 1$, the statement is true. $\quad 4^1 < 4^{1+1}$ ✓

Induction Step: Assume the statement is true for some $n = k$, where k is any natural number. This means $4^k < 4^{k+1}$. Since the statement is assumed true, we can use it to show the statement is true for $n = k + 1$.
Multiply both sides by 4. $\quad\quad\quad\quad\quad\quad\quad\quad\quad\quad 4 \cdot 4^k < 4 \cdot 4^{k+1}$
When multiplying, the exponents are added. $\quad\quad\quad\quad 4^{k+1} < 4^{k+2}$
Since S_k implies S_{k+1}, the statement $4^n < 4^{n+1}$ is true for all natural numbers by the principle of mathematical induction.

23. The statement we are proving true is $S_n : n < n+1$. The needed components for mathematical induction are $S_1 : 1 < 1+1$, $S_k : k < k+1$, and $S_{k+1} : k+1 < k+1+1$.

Basis Step: For $n = 1$, the statement is true. $\quad 1 < 1+1$ ✓

Induction Step: Assume the statement is true for some $n = k$, where k is any natural number. This means $k < k+1$. Since the statement is assumed true, we can use it to show the statement is true for $n = k + 1$.
Add 1 to both sides. $\quad\quad\quad\quad\quad\quad\quad\quad k+1 < k+1+1$

Since S_k implies S_{k+1}, the statement $n < n+1$ is true for all natural numbers by the principle of mathematical induction.

24. The statement we are proving true is $S_n : \log n < n$. The needed components for mathematical induction are $S_1 : \log 1 < 1$, $S_k : \log k < k$, and $S_{k+1} : \log(k+1) < k+1$.

Basis Step: For $n = 1$, the statement is true. $\quad \log 1 < 1$
$\quad\quad\quad\quad\quad\quad\quad\quad\quad\quad\quad\quad\quad\quad\quad\quad 0 < 1$ ✓

Induction Step: Assume the statement is true for some $n = k$, where k is any natural number. This means $\log k < k$. Since the statement is assumed true, we can use it to show the statement is true for $n = k + 1$.
Add 1 to both sides. $\quad\quad\quad\quad\quad\quad\quad 1 + \log k < 1 + k$
Replace 1 with $\log 10$. $\quad\quad\quad\quad\quad\quad \log 10 + \log k < 1 + k$
Condense logarithms using product rule. $\log(10k) < 1 + k$

Since $k + 1 < 10k$, then $\log(k+1) < \log(10k)$. Therefore, $\log(k+1) < \log(10k) < 1+k$, which means $\log(k+1) < k+1$.

Since S_k implies S_{k+1}, the statement $\log n < n$ is true for all natural numbers by the principle of mathematical induction.

25. The statement we are proving true is $S_n : 2n \leq 2^n$. The needed components for mathematical induction are $S_1 : 2(1) \leq 2^1$, $S_k : 2k \leq 2^k$, and $S_{k+1} : 2(k+1) \leq 2^{k+1}$.

Basis Step: For $n = 1$, the statement is true. $\quad 2(1) \leq 2^1$ ✓

Induction Step: Assume the statement is true for some $n = k$, where k is any natural number. This means $2k \leq 2^k$. Since the statement is assumed true, we can use it to show the statement is true for $n = k + 1$.
Multiply both sides by 2. $\quad\quad\quad\quad\quad\quad 2 \cdot 2k \leq 2 \cdot 2^k$
Since $1 \leq k$, then $k+k+1+1 \leq k+k+k+k$. Therefore, $2k+2 \leq 4k$, which means $2k+2 \leq 2 \cdot 2k \leq 2 \cdot 2^k$. It follows that $2(k+1) \leq 2^{k+1}$.

Since S_k implies S_{k+1}, the statement $2n \leq 2^n$ is true for all natural numbers by the principle of mathematical induction.

Lesson 23   Binomial Theorem

In combinatorics, the number of combinations of n objects taken r at a time is denoted by ${}_nC_r$.

$${}_nC_r = \frac{n!}{r!(n-r)!}$$

If there are 5 students in a classroom and 3 are chosen at random, then 10 different combinations are possible.

$${}_5C_3 = \frac{5!}{3!(5-3)!} = \frac{5 \cdot 4 \cdot 3 \cdot 2 \cdot 1}{3!2!} = 10$$

If you are packing t-shirts for a vacation and you need to take 7, how many different t-shirt combinations can you make if there are 15 shirts to choose from in your closet?

$${}_{15}C_7 = \frac{15!}{7!(15-7)!} = \frac{15!}{7!8!} = 6435$$

There are drink machines in restaurants that offer many choices and flavors. If there are 10 choices and you pick 2, how many combinations are available?

$${}_{10}C_2 = \frac{10!}{2!(10-2)!} = \frac{10 \cdot 9 \cdot 8!}{2!8!} = \frac{10 \cdot 9}{2} = 45$$

Note: It is impossible to choose 11 flavors if only 10 are available. Therefore, $r \le n$.

Example 1: Evaluate the following.
a. ${}_4C_0$ b. ${}_4C_1$ c. ${}_4C_2$ d. ${}_4C_3$ e. ${}_4C_4$

Solution:

a. ${}_4C_0 = \dfrac{4!}{0!(4-0)!} = 1$ b. ${}_4C_1 = \dfrac{4!}{1!(4-1)!} = 4$ c. ${}_4C_2 = \dfrac{4!}{2!(4-2)!} = 6$

d. ${}_4C_3 = \dfrac{4!}{3!(4-3)!} = 4$ e. ${}_4C_4 = \dfrac{4!}{4!(4-4)!} = 1$

Your calculator will compute the combination by pressing MATH and move the cursor to the PRB submenu.

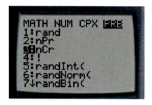
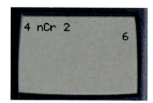

Combinatorics is a branch of mathematics used in finite math, calculus, and statistics. It can also be used here in pre-calculus. Take a look at $(a+b)^4$. To expand this binomial, we need to multiply $\underbrace{(a+b)(a+b)}(a+b)(a+b)$. Group the first two factors and FOIL.

$(a^2+2ab+b^2)(a^2+2ab+b^2)$ Distribute and combine like terms to find the expansion.

$a^4+4a^3b+6a^2b^2+4ab^3+b^4$

Do you notice any similarities between the answers in Example 1 and the coefficients of the terms in the expansion? The numbers 1, 4, 6, 4, and 1 are not a coincidence! The combination formula for $_nC_r$ produces the coefficients of the binomial expansion. Also, notice the exponents of the variables always add to 4. In fact, the powers of a decrease as the powers of b increase. All of these patterns are summarized in the Binomial Theorem.

The Binomial Theorem

For any binomial $a+b$ and any natural number n,

$$(a+b)^n = \binom{n}{0}a^n b^0 + \binom{n}{1}a^{n-1}b^1 + \binom{n}{2}a^{n-2}b^2 + \ldots + \binom{n}{n-1}a^1 b^{n-1} + \binom{n}{n}a^0 b^n$$

where $\binom{n}{r} = {_nC_r}$.

You can write the binomial theorem as a series using sigma notation. $(a+b)^n = \sum_{r=0}^{n}\binom{n}{r}a^{n-r}b^r$

Because the sigma notation begins with $r = 0$, there will be $n + 1$ terms in the binomial expansion.

Example 2: Expand the expression $(x-3)^5$ using the binomial theorem.

Solution:

$$(x-3)^5 = \sum_{r=0}^{5}\binom{5}{r}x^{5-r}(-3)^r$$

$$= \binom{5}{0}x^5(-3)^0 + \binom{5}{1}x^4(-3)^1 + \binom{5}{2}x^3(-3)^2 + \binom{5}{3}x^2(-3)^3 + \binom{5}{4}x^1(-3)^4 + \binom{5}{5}x^0(-3)^5$$

$$= 1x^5(1) + 5x^4(-3) + 10x^3(9) + 10x^2(-27) + 5x(81) + 1(1)(-243)$$

$$= x^5 - 15x^4 + 90x^3 - 270x^2 + 405x - 243$$

Notice, the expansion is an alternating series. The signs alternate between positive and negative terms.

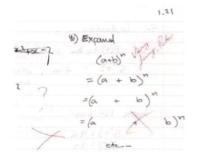

Don't be like this student… 😊
expanding a binomial does not mean to write it wider with more and more space between the variables.

There is another pattern associated with the coefficients.

$(a+b)^0 =$ 1

$(a+b)^1 =$ $a+b$

$(a+b)^2 =$ $a^2 + 2ab + b^2$

$(a+b)^3 =$ $a^3 + 3a^2b + 3ab^2 + b^3$

$(a+b)^4 =$ $a^4 + 4a^3b + 6a^2b^2 + 4ab^3 + b^4$

```
            1
          1   1
        1   2   1
      1   3   3   1
    1   4   6   4   1
  ?   ?   ?   ?   ?   ?
```

Can you guess the next row of numbers?

There are 1's on the outsides, and the middle numbers are the sum of the two numbers above it.

This triangular array of numbers is called **Pascal's triangle**. We can expand $(a+b)^5$ using the triangle to determine the coefficients. Remember, the powers of a decrease as the powers of b increase. The sum of the exponents will equal 5.

$$(a+b)^5 = a^5 + 5a^4b + 10a^3b^2 + 10a^2b^3 + 5ab^4 + b^5$$

Example 3: Expand the expression $(5x + 2y)^6$ using the binomial theorem.

Solution: Using the 7th row of Pascal's triangle, we can find the coefficients.

$$(5x+2y)^6 = \sum_{r=0}^{6}\binom{6}{r}(5x)^{6-r}(2y)^r$$

$$= \underline{1}(5x)^6 + \underline{6}(5x)^5(2y)^1 + \underline{15}(5x)^4(2y)^2 + \underline{20}(5x)^3(2y)^3 + \underline{15}(5x)^2(2y)^4 + \underline{6}(5x)^1(2y)^5 + \underline{1}(2y)^6$$

$$= 15625x^6 + 37500x^5y + 37500x^4y^2 + 20000x^3y^3 + 6000x^2y^4 + 960xy^5 + 64y^6$$

One common mistake when expanding the binomial is forgetting to apply the exponent to the number and the variable. Remember: $(5x)^6 = 5^6 x^6$.

In algebra, you learned how to multiply complex numbers using FOIL. This method works great for two binomials. $(4-3i)^2 = (4-3i)(4-3i)$

$= 16 - 12i - 12i + 9i^2$ ← Reminder: $i^2 = -1$

$= 16 - 24i + 9(-1)$

$= 7 - 24i$

Powers of i

$i^0 = 1$	$i^4 = 1$
$i^1 = i$	$i^5 = i$
$i^2 = -1$	$i^6 = -1$
$i^3 = -i$	$i^7 = -i$

For higher exponents, the binomial theorem provides a more efficient method to multiply the binomial.

Example 4: Expand the complex number $(4-3i)^4$ using the binomial theorem.

Solution: Using the 5th row of Pascal's triangle, we can find the coefficients.

$(4-3i)^4 = \underline{1}(4)^4(-3i)^0 + \underline{4}(4)^3(-3i)^1 + \underline{6}(4)^2(-3i)^2 + \underline{4}(4)^1(-3i)^3 + \underline{1}(4)^0(-3i)^4$

$= 256 - 768i + 864i^2 - 432i^3 + 81i^4$

$= 256 - 768i + 864(-1) - 432(-i) + 81(1)$

$= -527 - 336i$

Suppose you need to find only the 5th term. To find the entire binomial expansion would be inefficient. If we are using the Binomial Theorem $(a+b)^n = \sum_{r=0}^{n}\binom{n}{r}a^{n-r}b^r$, we can find any one term without having to find the entire expansion. Notice the coefficient $\binom{n}{r}$ starts with $r = 0$ for the first term. If we are looking for the 5th term, then $r = 4$. Notice the exponent of b is equal to r. If we are looking for the 5th term, then b is raised the 4th power. We can use the following formula to find any term of the expansion. Always remember the exponents will add to n.

> **Finding a Specific Term of a Binomial Expansion**
>
> For the binomial expansion $(a+b)^n$, the kth term is given by
>
> $$\binom{n}{k-1} a^{n-k+1} b^{k-1}.$$

Example 5: Find the indicated term for each binomial expression.

a. 5th; $(x+y)^8$ b. 2nd; $(2x-5y)^6$

Solution:

a. $n = 8$ and $k - 1 = 4$

$$\binom{8}{4} x^4 y^4 = 70 x^4 y^4$$

b. $n = 6$ and $k - 1 = 1$

$$\binom{6}{1}(2x)^5(-5y)^1 = 6(32x^5)(-5y) = -960 x^5 y$$

If the degree n is relatively large, don't forget the formula for finding the coefficients using combinatorics.

$$_nC_r = \frac{n!}{r!(n-r)!}$$

Example 6: Find the 10th term for binomial expression $\left(\dfrac{3}{x} + y\right)^{13}$.

Solution: $n = 13$ and $k - 1 = 9$

$$\binom{13}{9}\left(\frac{3}{x}\right)^4 (y)^9 = \frac{13!}{9!(13-9)!}\left(\frac{81}{x^4}\right) y^9 = \frac{13(12)(11)(10)9!}{9!4!}\left(\frac{81 y^9}{x^4}\right) = \frac{17160}{24}\left(\frac{81 y^9}{x^4}\right) = \frac{57915 y^9}{x^4}$$

Lesson 23 — Practice Exercises

In Exercises 1 – 4, evaluate the following.

1. $\binom{6}{3}$
2. $\binom{7}{0}$
3. $\binom{5}{2}$
4. $\binom{9}{9}$

In Exercises 5 – 12, use the binomial theorem to expand each binomial expression.

5. $(x+2)^4$
6. $(2x-1)^5$
7. $\left(\dfrac{1}{x}+6y\right)^3$
8. $(x^2+\sqrt{6})^6$

9. $(x^3-8)^7$
10. $(4x-5y)^8$
11. $(1-2i)^5$
12. $(3+5i)^6$

In Exercises 13 – 10, find the indicated term for each binomial expression.

13. 4th; $(x-3y)^7$
14. 7th; $(x^2+y^2)^{10}$
15. 5th; $(2x+y)^9$

16. 8th; $(4x-\sqrt{2})^8$
17. 2nd; $\left(\dfrac{1}{x}+y\right)^3$
18. 9th; $(x^3-1)^{12}$

Solutions to Practice Exercises — Lesson 23

1. 20
2. 1
3. 10
4. 1

5. $x^4+8x^3+24x^2+32x+16$
6. $32x^5-80x^4+80x^3-40x^2+10x-1$

7. $\dfrac{1}{x^3}+\dfrac{18y}{x^2}+\dfrac{108y^2}{x}+216y^3$
8. $x^{12}+6x^{10}\sqrt{6}+90x^8+120x^6\sqrt{6}+540x^4+216x^2\sqrt{6}+216$

9. $x^{21}-56x^{18}+1344x^{15}-17920x^{12}+143360x^9-688128x^6+1835008x^3-2097152$

10. $65536x^8-655360x^7y+2867200x^6y^2-7168000x^5y^3+11200000x^4y^4-11200000x^3y^5+7000000x^2y^6-2500000xy^7+390625y^8$

11. $41+38i$
12. $39104-3960i$
13. $-945x^4y^3$
14. $210x^8y^{12}$

15. $4032x^5y^4$
16. $-256x\sqrt{2}$
17. $\dfrac{3y}{x^2}$
18. $495x^{12}$

Cumulative Review 5   Lessons 19 – 23

1. Write the first five terms of each sequence.

   a. $a_n = 4n + 10$
   b. $a_n = 2^{n-1}$
   c. $a_n = \dfrac{4n-1}{n^2+2}$

   d. $a_n = (-1)^{n-1}(n+1)$
   e. $a_1 = -2$, $a_n = a_{n-1} + 3$
   f. $a_1 = 2$, $a_2 = 5$, $a_n = a_{n-1} + a_{n-2}$

2. Evaluate each series.

   a. $\sum_{i=1}^{5} 2i + 1$
   b. $\sum_{k=1}^{6} (-1)^k k$
   c. $\sum_{j=1}^{4} \dfrac{1}{j}$
   d. $\sum_{i=1}^{5} (i+1)^{-1}$

3. Use summation notation to write each series.

   a. $\dfrac{1}{3} + \dfrac{1}{6} + \dfrac{1}{9} + \ldots + \dfrac{1}{27}$
   b. $1 - \dfrac{1}{2} + \dfrac{1}{4} - \dfrac{1}{8} + \ldots - \dfrac{1}{128}$

4. Determine if the sequence is arithmetic, geometric, or neither. The find the general term, a_n.

   a. $-8, -12, -16, -20, \ldots$
   b. $-4, -12, -36, -108, \ldots$
   c. $2, 6, 12, 20, \ldots$

   d. $\dfrac{1}{2}, \dfrac{1}{3}, \dfrac{1}{4}, \dfrac{1}{5}, \ldots$
   e. $1, 4, 7, 10, 13, \ldots$
   f. $\dfrac{3}{4}, 1, \dfrac{4}{3}, \dfrac{16}{9}, \dfrac{64}{27}, \ldots$

5. Find a_1 and d for the arithmetic sequence when $a_5 = 27$ and $a_{15} = 87$.

6. Determine the number of terms in the sequence $-8, -5, -2, \ldots, 109$.

7. Find the partial sum S_{10} of the arithmetic sequence. $8, 11, 14, \ldots$

8. Find a_5 and a_n if $a_4 = 18$ and $r = 2$ for the geometric sequence.

9. Find a_1 and r for the geometric sequence if $a_2 = -6$ and $a_7 = -192$.

10. Find the partial sum S_5 and S_∞ (if possible).

    a. $2, 8, 32, 128, \ldots$
    b. $18, -9, \dfrac{9}{2}, \dfrac{-9}{4}, \ldots$

11. Five years ago, the population of a city was 49,000. Each year the zoning commission permits an increase of 580 in the population. What will the maximum population be 5 year from now?

12. Each person has two parents, four grandparents, eight great-grandparents, and so on. What is the total number of ancestors a person has, going back five generations? ten generations?

13. Prove the following statements using mathematical induction.

   a. $3+6+9+...+3n = \dfrac{3n(n+1)}{2}$

   b. $1+3+5+...+2n-1 = n^2$

14. Expand the binomial using the Binomial Theorem.

   a. $(x+y)^6$

   b. $(7p-2q)^4$

15. Find the 6th term of $(4h-j)^8$.

16. Find the 5th term of $(2x+5y)^6$.

Solutions to Review

1a. 14, 18, 22, 26, 30

1b. 1, 2, 4, 8, 16

1c. $1, \dfrac{7}{6}, 1, \dfrac{5}{6}, \dfrac{19}{27}$

1d. 2, -3, 4, -5, 6

1e. -2, 1, 4, 7, 10

1f. 2, 5, 7, 12, 19

2a. 35

2b. 3

2c. $\dfrac{25}{12}$

2d. $\dfrac{29}{20}$

3a. $\sum_{k=1}^{9} \dfrac{1}{3k}$

3b. $\sum_{k=1}^{8} (-1)^{k-1} \dfrac{1}{2^{k-1}}$

4a. arithmetic, $a_n = -8 + (n-1)(-4) = -4n - 4$

4b. geometric, $a_n = (-4)(3)^{n-1}$

4c. neither, $a_n = n(n+1)$

4d. neither, $a_n = \dfrac{1}{n+1}$

4e. arithmetic, $a_n = 1 + (n-1)3 = 3n - 2$

4f. geometric, $a_n = \dfrac{3}{4}\left(\dfrac{4}{3}\right)^{n-1}$

5. $a_1 = 3, d = 6$

6. 40

7. 215

8. $a_5 = 36, a_n = \dfrac{9}{4}(2)^{n-1}$

9. $a_1 = -3, r = 2$

10a. $S_5 = 682; S_\infty$ is not possible

10b. $S_5 = \dfrac{99}{8}; S_\infty = 12$

11. 54,800

12. 62; 2046

13a. The statement we are proving true is $3+6+9+...+3n = \dfrac{3n(n+1)}{2}$. The needed components for mathematical induction are $S_1: 3 = \dfrac{3(1)(1+1)}{2}$, $S_k: 3+6+9+...+3k = \dfrac{3k(k+1)}{2}$, and $S_{k+1}: 3+6+9+...+3k+3(k+1) = \dfrac{3(k+1)(k+1+1)}{2}$.

Basis Step: For $n = 1$, the statement is true. $S_1 = a_1$

$\dfrac{3(1)(1+1)}{2} = 3$

Induction Step: Assume the statement is true for some $n = k$, where k is any natural number. This means $S_k = \dfrac{3k(k+1)}{2}$. Verify the statement is true for $n = k + 1$. In other words, show $S_{k+1} = \dfrac{3(k+1)(k+1+1)}{2} = \dfrac{3(k+1)(k+2)}{2}$.

$S_{k+1} = 3+6+9+...+3k+3(k+1)$

$= S_k + a_{k+1}$

$= \dfrac{3k(k+1)}{2} + 3(k+1)$

$= \dfrac{3k(k+1)}{2} + \dfrac{2 \cdot 3(k+1)}{2}$

$= \dfrac{3k(k+1) + 6(k+1)}{2}$

250

$$= \frac{3(k+1)(k+2)}{2} \checkmark$$

Since S_k implies S_{k+1}, the statement $3+6+9+...+3n = \frac{3n(n+1)}{2}$ is true for all natural numbers by the principle of mathematical induction.

13b. The statement we are proving true is $1+3+5+...+(2n-1) = n^2$. The needed components for mathematical induction are $S_1: 1=1^2$, $S_k: 1+3+5+...+(2k-1) = k^2$, and $S_{k+1}: 1+3+5+...+(2k-1)+[2(k+1)-1] = (k+1)^2$.

Basis Step: For $n = 1$, the statement is true. $S_1 = a_1$
$$1^2 = 1$$

Induction Step: Assume the statement is true for some $n = k$, where k is any natural number. This means $S_k = k^2$. Verify the statement is true for $n = k+1$. In other words, show $S_{k+1} = (k+1)^2$.

$$S_{k+1} = 1+3+5+...+(2k-1)+[2(k+1)-1]$$
$$= S_k + a_{k+1}$$
$$= k^2 + 2(k+1) - 1$$
$$= k^2 + 2k + 2 - 1$$
$$= (k+1)(k+1)$$
$$= (k+1)^2 \checkmark$$

Since S_k implies S_{k+1}, the statement $1+3+5+...+(2n-1) = n^2$ is true for all natural numbers by the principle of mathematical induction.

14a. $x^6 + 6x^5y + 15x^4y^2 + 20x^3y^3 + 15x^2y^4 + 6xy^5 + y^6$

14b. $2401p^4 - 2744p^3q + 1176p^2q^2 - 224pq^3 + 16q^4$

15. $-3584h^3j^5$ **16.** $37{,}500x^2y^4$

Just for Fun….because math is fun!!

If you multiply the following string of binomials, what is the result?

$$(x-a)(x-b)(x-c)(x-d)...(x-y)(x-z)$$

ALGEBRA REVIEW

Lesson 1 Page 13

1. $2(6x^2 + 17x + 12)$
2. $6x(x+4)(x-5)$
3. $(5x - 7y)^2$
4. $(x+2)(x-2)(x+3)(x-3)$
5. $3(x^2 + 8)(x^2 - 8)$
6. $(x^2 + 4)(3x - 1)$
7. $2(3x - 4y)(9x^2 - 12xy + 16y^2)$
8. $(2x + 5y)(4x^2 - 10xy + 25y^2)$
9. $x^{1/3}(x+1)(x-1)$
10. $\dfrac{5(x+3)}{x^{1/2}}$

Lesson 2 Page 25

1a. $3x^2 + 2x - 2 - \dfrac{2x - 6}{x^2 + 1}$

1b. $2x^2 + 6 + \dfrac{11x^2 + 18x + 36}{x^3 - 3x - 4}$

1c. $-3x^4 - x^3 + 3x^2 - 1 - \dfrac{x - 6}{3x^2 - x + 1}$

2a. $\{-2, 8\}$

2b. $\left\{\dfrac{4 - 4\sqrt{2}}{3}, \dfrac{4 + 4\sqrt{2}}{3}\right\}$

2c. $\{0, 3\}$

2d. $\{\pm 3i\}$

2e. $\{-3, 7\}$

2f. $\left\{\dfrac{6 - i\sqrt{3}}{3}, \dfrac{6 + i\sqrt{3}}{3}\right\}$

Lesson 4 Page 45

1. $\{-2 \pm 2\sqrt{6}\}$

2. $\left\{\dfrac{3 - i\sqrt{71}}{4}, \dfrac{3 + i\sqrt{71}}{4}\right\}$

3. $\sqrt{58}$

4. $\dfrac{\sqrt{37}}{2}$

5. $\left(-\dfrac{3}{2}, \dfrac{3}{2}\right)$

6. $\left(-1, \dfrac{1}{2}\right)$

Lesson 5 Page 61

1. $(-\infty, -2)$
2. $(-\infty, 36]$
3. $(-9, 6)$

4. $[-3,3]$ 5. $\left(-\infty,-\dfrac{5}{3}\right]\cup\left[\dfrac{5}{3},\infty\right)$ 6. $\left(-\infty,-\dfrac{19}{2}\right)\cup\left(-\dfrac{5}{2},\infty\right)$

Lesson 6 Page 73

1. $\dfrac{2x-6}{x+2}=\dfrac{2(x-3)}{x+2}$ 2. $\dfrac{x-2}{x-4}$ 3. $\dfrac{3x-1}{(x+1)(x-1)}$

4. $\dfrac{x^2+1}{x(x+1)^2}$ 5. $\dfrac{x-11}{(x+4)(x-2)(x-1)}$ 6. $\dfrac{x-x^2}{(x-1)^2}=\dfrac{-x}{x-1}$

Lesson 7 Page 87

1. $f(g(x))=3x^2-2$; $g(f(x))=9x^2-30x+26$

2. $f(g(x))=\sqrt{4x+3}$; $g(f(x))=3+4\sqrt{x}$

3. $f(g(x))=x^2-1$; $g(f(x))=x^2+2x-1$

4. $f^{-1}(x)=\dfrac{x+1}{2}$

	Domain	Range
$f(x)$	$(-\infty,\infty)$	$(-\infty,\infty)$
$f^{-1}(x)$	$(-\infty,\infty)$	$(-\infty,\infty)$

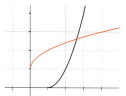

5. $f^{-1}(x)=\sqrt{x}+1$

	Domain	Range
$f(x)$	$[1,\infty)$	$[0,\infty)$
$f^{-1}(x)$	$[0,\infty)$	$[1,\infty)$

6. $f^{-1}(x)=(x-3)^3$

	Domain	Range
$f(x)$	$(-\infty,\infty)$	$(-\infty,\infty)$
$f^{-1}(x)$	$(-\infty,\infty)$	$(-\infty,\infty)$

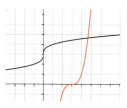

Lesson 9 Page 102

1. 3 billion   2. 7.7 billion   3. 98.5 years; between 2058 and 2059

4. 11.778 billion

KEEP IT FRESH

Lesson 6 Page 77

1. $f(-4) = -1$ and $f(3) = -3$.

2.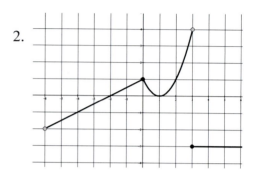

3. $(-6, \infty)$ 4. $\{-3\} \cup (-2, 4)$

5. Continuous at $x = 0$. Discont at $x = 3$.

Lesson 13 Page 144

1. $(3x+2)(2x-1)(x-2)^2$

2. $P(0) = -3$ and $P(1) = 2$; Since $P(0)$ and $P(1)$ have opposite signs, there is at least one zero in the interval.

3a. $\pm 1, \pm \dfrac{1}{2}, \pm 2, \pm 3, \pm \dfrac{3}{2}, \pm 6$

3b.

P	N	C
2	1	0
0	1	2

3c. $\left\{-\dfrac{1}{2}, 2, 3\right\}$

3d. Far Left: down/$-\infty$; Far Right: up/∞

3e.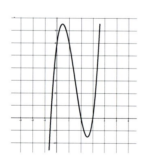

4. $f(x) = \begin{cases} -3x + 4 & \text{for } x \leq 4/3 \\ 3x - 4 & \text{for } x > 4/3 \end{cases}$

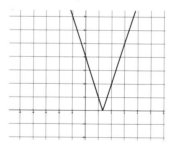

Lesson 13 Page 146

1a. $(-2, 1) \cup (2, \infty)$

1b. $(-\infty, -3] \cup \left[-\dfrac{1}{2}, \dfrac{1}{3}\right]$

2a. $(-\infty,-3)\cup[-2,3)$ 2b. $(-\infty,-3)\cup\left[-\dfrac{1}{5},1\right)$

Lesson 14 Page 153

1a. $\log\dfrac{x^2 y^3}{\sqrt{z}}$ 1b. $\ln x(x-2)$

2a. $\dfrac{1}{2}(2+\log x)$ 2b. $1-3\ln y$

3a. Left 3 and Down 5   3b. Left 3 and Reflect x-axis

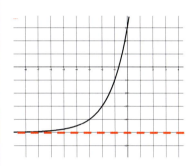 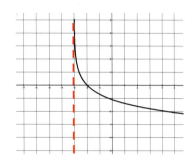

Lesson 18 Page 187

1a. $(-\infty,3)\cup(3,\infty)$ 1b. $x=3$ 1c. $y=x+4$ 1d. $(-2,0)\,\&\,(1,0)$

1e. $\left(0,\dfrac{2}{3}\right)$ 1g. 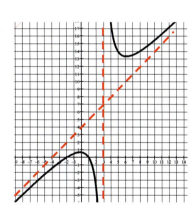   2. $(1,1)$

Lesson 19 Page 203

1. $\begin{bmatrix} -1 & 3 & 2 & | & -10 \\ 3 & -2 & -2 & | & 7 \\ -2 & 1 & -1 & | & -10 \end{bmatrix}$; $(1,-5,3)$

2a. $\dfrac{2}{x-1}+\dfrac{1}{x+3}+\dfrac{3}{(x+3)^2}$ 2b. $\dfrac{-1}{x+1}+\dfrac{2x-3}{x^2+4}$

Lesson 21 Page 224

1. 2×2, 2×4, 2×2

2. $\begin{bmatrix} 8 & 3 \\ -3 & 0 \end{bmatrix}$

3. $\begin{bmatrix} -1 & -2 & -3 & -4 \\ 0 & -2 & 1 & -6 \end{bmatrix}$

4. $\begin{bmatrix} 4 & 12 & 10 & 28 \\ 1 & 8 & 0 & 22 \end{bmatrix}$

5. -5

6. $C^{-1} = \begin{bmatrix} -6/5 & 1/5 \\ 1 & 0 \end{bmatrix}$

7. $AX = B \qquad X = A^{-1}B$

$\begin{bmatrix} 1 & 2 & 5 \\ 2 & 3 & 8 \\ -1 & 1 & 2 \end{bmatrix} \begin{bmatrix} x \\ y \\ z \end{bmatrix} = \begin{bmatrix} 2 \\ 3 \\ 3 \end{bmatrix}$

$\begin{bmatrix} x \\ y \\ z \end{bmatrix} = \begin{bmatrix} 2 & -1 & -1 \\ 12 & -7 & -2 \\ -5 & 3 & 1 \end{bmatrix} \begin{bmatrix} 2 \\ 3 \\ 3 \end{bmatrix} = \begin{bmatrix} -2 \\ -3 \\ 2 \end{bmatrix}$

8. $(5, 2, 0)$

9. 17 sq units

Lesson 21 Page 228

1.
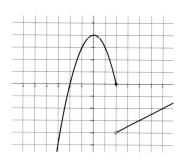

2.
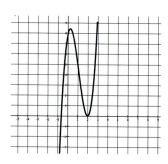

3.
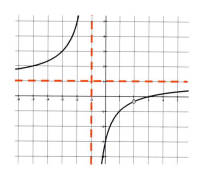

4.
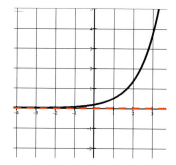

5.
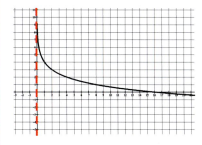

6.
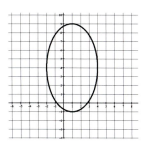

7.
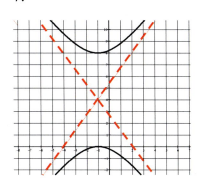

8.
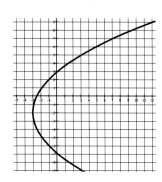

Lesson 22 Page 234

1. $\dfrac{(x-1)^2}{4} + \dfrac{(y+2)^2}{9} = 1$; $(1,-2)$; $(1,-5)\,\&\,(1,1)$; $(1,-2\pm\sqrt{5})$

2. $(2,-1)$; $(-1,-1)\,\&\,(5,-1)$; $(-3,-1)\,\&\,(7,-1)$; $y+1 = \pm\dfrac{4}{3}(x-2)$

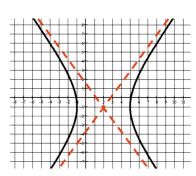

3. $y^2 = 20x$

Math is Fun: If you multiply the following string of binomials, what is the result? $(x-a)(x-b)(x-c)(x-d)...(x-y)(x-z)$

Answer: 0, because $(x-x) = 0$.

Index

Absolute Value Function	11	Factor Theorem	22
Additive Identity	119	Factorial	201
Additive Inverse	118	Fibonacci Sequence	198
Alternating Sequence	194	Finite Sequence	194
Area of a Triangle	142	Finite Series	198
Arithmetic Sequence	208	Focal Chord	182
Arithmetic Series	212	Foci	164,172
Augmented Matrix	106	Focus	181
Axis of Symmetry	182	Gaussian Elimination	107
Basis Step	231	Gauss-Jordan Elimination	111
Binomial Theorem	244	General Term	193,209,218
Central Rectangle	173	Geometric Sequence	216
Change of Base Formula	91	Geometric Series	221,222
Circle	164	Greatest Common Factor	158
Coefficient Matrix	124, 131	Greatest Integer Function	8
Cofactor	137	Horizontal Asymptote	55
Collinear	143	Hyperbola	172
Column Matrix	131	Identity Matrix	127
Column of a Matrix	106	Imaginary Numbers	246
Column Rotation	139	Induction Step	231
Combination	243	Infinite Geometric Series	222
Common Difference	208	Infinite Sequence	194
Common Ratio	216	Infinite Series	198
Complex Numbers	29,246	Intermediate Value Theorem	24
Conic Sections	164	Inverse Functions	78
Conjugate	29	Inverse Matrix	127,136
Conjugate Axis	172	Invertible Matrix	130
Convergent	222	Latus Rectum	182
Cramer's Rule	140	Leading Term Test	39
Decartes Rule of Signs	32	Limit	222
Determinant	136	Logarithmic Equations	97
Difference of Cubes	158	Logarithmic Function	78
Difference of Squares	158	Long Division	20,58,152
Directrix	181	Major Axis	165
Distance Formula	45	Mathematical Induction	229,231
Divergent	222	Matrix	106
Eccentricity	165	Matrix Multiplication	121
Ellipse	164	Midpoint Formula	45
Entry of a Matrix	106	Minor	137
Exponential Equations	95	Minor Expansion	137
Exponential Function	78	Multiplicity	29,42
Extraneous Solution	100	Nonsingular Matrix	130

Oblique Asymptote	55
One-to-One	78
Order of a Matrix	106
Parabola	181
Partial Fraction Decomposition	147
Pascal's Triangle	245
Perfect Square Trinomial	158
Piecewise Function	6
Polynomial	20,29,39
Polynomial Inequalities	66
Power Rule	89
Product Rule	88
Properties of Matrices	123
Properties of Summation	201
Quadratic Formula	31
Quotient Rule	89
Radius	164
Rational Function	54
Rational Inequalities	70
Rational Zeros Theorem	30
Recursion Formula	196
Recursively Defined Sequence	196
Reduced Row-Echelon Form	110
Relative Extrema	40
Remainder Theorem	21
Removable Discontinuity	8,54
Row of a Matrix	106
Row Operations	107
Row-Echelon Form	107
Sequence	193
Series	198
Sigma Notation	199
Singular Matrix	130
Slant Asymptote	55
Sum of Cubes	158
Summation Notation	199
Synthetic Division	20
Transformations	6,10,81
Transverse Axis	172
Vertex	181
Vertical Asymptote	54
Vertices	165,172
Zero	23,29

Photo Credits

Lesson 4:
Page 40 (middle) TI-84 Plus/Texas Instruments
Page 50 (bottom) Millennium Force/Creative Commons

Lesson 7:
Page 82 (middle) TI-84 Plus/Texas Instruments
Page 83 (bottom) Log and Ln/Chris Burke/xwhy? Used by permission

Lesson 13:
Page 144 (top) TI-84 Plus/Texas Instruments
Page 144 (bottom) Final Exam/Clipart Finders

Lesson 14:
Page 154 (bottom) Warning/Amazon Sign

Lesson 16:
Page 164 (top) Conics/2012 books by Lardbucket
Page 168 (middle) Archway/Clipart

Lesson 17:
Page 172 (top) Conics/2012 books by Lardbucket
Page 177 (middle) Nuclear Tower/Creative Commons

Lesson 18:
Page 181 (top) Conics/2012 books by Lardbucket
Page 186 (bottom) Golden Gate Bridge/Bedford

Lesson 22:
Page 233 (top) Dominoes/Creative Commons
Page 242 (bottom) Dominoes/Creative Commons

Made in United States
Orlando, FL
07 December 2022